DIALOGUE ON A MOUNTAINTOP:

Surveyor Laommi Baldwin: **"Mr. Crocker, all I am doing is to follow the contour of these hills, and God arranged the contour."**

Railroad promoter Alvah Crocker: **"Well, if God did that much, why didn't He poke His thumb through this darned mountain and save us all the trouble of trying to make a tunnel."**

Beginning with this moment of truth in 1826, A PINPRICK OF LIGHT relates the story of the Troy & Greenfield RR and its Hoosac Tunnel in greater depth and detail than ever before—all the big trouble it was to build and some unforeseen troubles in running trains through it ever since!

When completed, Hoosac was *the second longest railroad tunnel in the world*, piercing a mountain almost as thick as Everest is high, and a pioneer engineering feat resulting in "a new rule book for both mining and politicking," the latter by Alvah Crocker, who had to mastermind the money out of the Massachusetts legislature in the face of skilled and unprincipled opposition.

But without Hoosac, Crocker's Troy & Greenfield "couldn't get there from here." The trying to make a tunnel began:

Chief Engineer Edwards tried, with a boring machine that conked out after chewing a hole twelve feet deep.

The redoubtable Herman Haupt tried, with oratory, hand drills and black powder, then went broke and quit to join the Union Army.

The state of Massachusetts tried. A lethal disaster at Central Shaft brought pressure to abandon the "bloody pit" once and for all.

It took a contractor from Canada and a chemist from England to save the project, using those new-fangled pneumatic drills and tri-nitroglycerin explosive, an historic "first time" for these methods in America.

Ironically, Alvah Crocker never lived to ride a train through the "Great Bore" of which he was the prime mover. But millions of others have, and it is one of the pleasures of this book to join the author in the cab of a Boston & Maine diesel over the route from Greenfield into the enveloping inky blackness of Hoosac Tunnel, unrelieved except for—miles ahead—a pinprick of light.

A PINPRICK OF LIGHT

Cross section of Hoosac Mountain, looking north, showing both West and Central Shafts. Not shown are four other shafts located between West Portal and West Shaft. From an 1880 post card. *Ruth B. Browne collection, courtesy of North Adams Public Library*

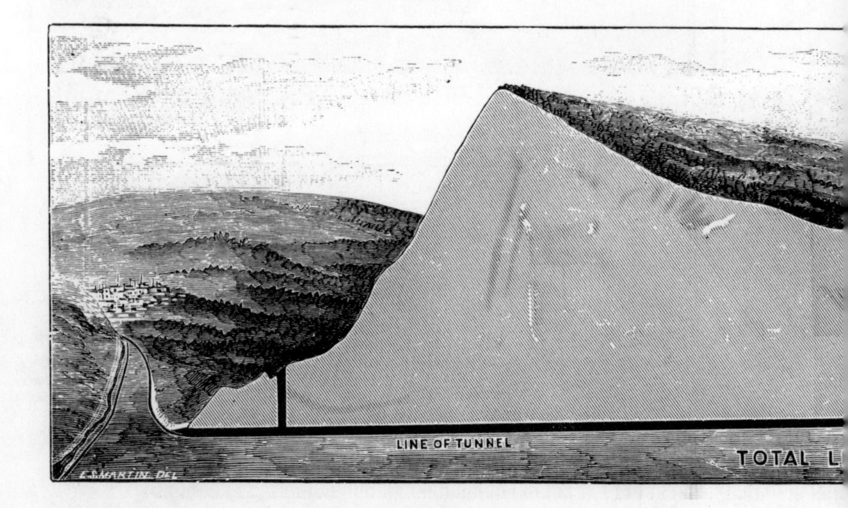

A PINPRICK OF LIGHT

The Troy and Greenfield Railroad and Its

HOOSAC TUNNEL by CARL R. BYRON

THE STEPHEN GREENE PRESS Brattleboro, Vermont

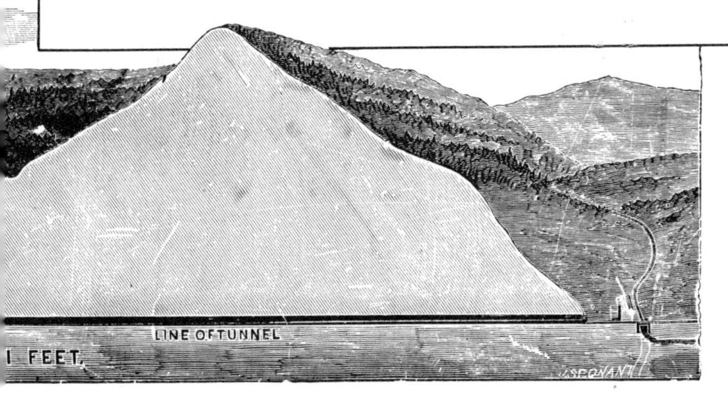

ACKNOWLEDGEMENT

The following individuals are just a few of many who have helped create *A Pinprick of Light* from their records, memoirs, photographs or family histories:

Mr. Randy Trabold, *North Adams Transcript;* Mr. John W. Barriger,* Chief Executive Officer, Boston and Maine Corporation; Mr. Richard Symmes, Beverly (Mass.) Historical Society; Mrs. Sondra Burnes, New England Power Company;

Mrs. Biglow Crocker; Mr. William Lasher*; Mr. William Fletcher: Mr. William Middleton; Mrs. Percy Bugbee*; Mr. Edward Ellis; Mr. Win Nowell; Mr. Don Barbeau, and others too numerous to enumerate here. A special note of appreciation must go to Russ Hamilton for the format design of this edition.

Thank you everyone.

**Deceased*

Originally published in altered form by the author, in a limited, hardbound edition with the imprint of Carlton Press, Inc.

This book has been produced in the United States of America. It is designed by R. Dike Hamilton and published by The Stephen Greene Press, Brattleboro, Vermont 05301.

Library of Congress Cataloging in Publication Data

Byron, Carl R 1948-
A pinprick of light

(Shortline RR series)
1. Hoosac Tunnel, Mass. 2. Boston and Maine Railroad. I. Title
TF238.H7B97 385'.312'097441 77-92774
ISBN 0-8289-0324-7

Cover photo by Peter Trabold, courtesy of Randy Trabold, *North Adams Transcript*

A PINPRICK OF LIGHT

1819: *First proposal for a canal through Hoosac between Boston and Albany, N.Y.* **1826:** *Canal proposal killed in Legislature.* **1842:** *Fitchburg RR from Boston to Fitchburg chartered. Opened 1845.* **1844:** *Vermont & Massachusetts RR chartered, Fitchburg to Brattleboro, Vt., with Greenfield spur; opened 1849.* **1848:** *Troy & Greenfield RR chartered between Greenfield and Vermont border at Williamstown.* **1849:** *Troy & Boston RR chartered between Troy, N.Y., and Vermont state line; opened 1852.* **1851:** *T&G construction in North Adams started.* **1852:** *Tunnel construction begun at East End (rock borer).* **1855:** *T&G signs contract with E. W. Serrell & Co.* **1856:** *Vermont Southern RR completed as link between T&G and T&B.* **1856:** *T&B starts North Adams–Troy service via the VS and T&G.* **1856:** *T&G annuls the Serrell contract, but negotiates a new one with H. Haupt & Co.* **1858:** *T&G renegotiates Haupt contract to include all tunnel and right-of-way construction.* **1860:** *Massachusetts Legislature amends the 1854 "Tunnel Aid" bill, and drops the $600,000 stock subscription requirement.*

CHAPTER  THE BEGINNING: 1819–1860

Placed squarely between Vermont's Green Mountains and the Appalachian Chain stands the Housatonic Range. These hills bisect Massachusetts from north to south and sever the Hudson and Connecticut River watersheds. Mt. Graylock is the highest of them, but Hoosac Mountain, twin-peaked and five miles wide at its base, is their kingpin.

Hoosac Mountain is primarily composed of limestone, mica, slate with some quartz veins, and gneiss; a little feldspar is also present. The center of the mountain is tough gneiss changing to talcose slate toward the east. Working westward from the center is mica, more slate marbled with quartz, feldspar, and a thin layer of limestone at the extreme western end.

As early as 1819, America's canal-building rage prompted a proposal to tunnel through Hoosac as part of a Boston-Troy—or Albany N.Y. canal. The project's magnitude quickly overwhelmed even its supporters' enthusiasm, and it quietly died. The towns of Worcester and Springfield resurrected the project in 1825 when they drummed up enough support to have a legislative "Commission" formed to study statewide transportation. The Commission hired civil engineer Loammi Baldwin (of Baldwin apple fame) to survey potential routes to "The West," and soon his crews were doing just that.

Baldwin's 1826 report to the legislature favored a northern route through Fitchburg and Greenfield rather than one through Worcester and Springfield due to better river conditions. The only roadblock to the northern route was the 2566-foot bulk of Hoosac—no tunnel, no route.

Even as the mountain overshadowed him, Baldwin's perseverance for this route led to the following exchange with fellow promoter Alvah Crocker as they stood on the rocky barrier:

> Baldwin: Mr. Crocker, all I am really doing is to follow the contour of these hills, and God arranged that contour.
>
> Crocker: Well, if God did that much, why didn't He poke His thumb through this darned mountain and save us all the trouble of trying to make a tunnel?

Baldwin felt that the entire tunnel and canal could be constructed for between $370,000 and $920,832. Legislative committee member Abner Phelps proposed to build a horsepowered "Rail-Road" over Baldwin's southern route instead of the canal and tunnel. However, the shaken legislators thought the "Rail-Road" too new, the canal unprofitable, and both overly expensive. Hence they paid Baldwin for his labors and killed both proposals.

ALVAH CROCKER *Courtesy of Mrs. Biglow Crocker*

The Western Railroad from Boston to Albany (N.Y.)—via Springfield and Worcester—opened for traffic in December, 1841, but left much to be desired according to many Northern and Western Massachusetts residents. The line usually didn't reach near them; and when it did, they had to pay outrageous rates to use it. Even more galling was the realization that their taxes had been used to finance its construction. These townspeople demanded equal treatment from the legislature.

Most prominent of these "protesters" was Alvah Crocker, the major paper manufacturer in Fitchburg, Massachusetts.

Born in 1801, he was a self-made man, having risen from a child laborer to successful businessman and highly respected Fitchburg citizen. He realized that the only way Fitchburg industry could compete with that near the Western R.R. was to likewise have a railroad outlet. Thus he left Crocker-Burbank Paper Co. in his partners' hands and with a group of fellow businessmen formed the Fitchburg R.R. during November, 1841.

Using his political know-how, Crocker wrestled a charter for the new railroad on March 2, 1842, in spite of Western R.R. opposition in the legislature. Construction from Boston to Fitchburg started immediately.

Picking up where the Fitchburg R.R. left off, in 1844 Crocker and associates incorporated the Vermont & Massachusetts R.R. from Fitchburg to Grout's Corner (now Millers Falls), and northwestward to Brattleboro, Vermont. A spur branched into Greenfield to connect with the North-South Connecticut River R.R.

The Fitchburg line was opened on March 5, 1845, with President Alvah Crocker riding on the first train from Boston. Despite the achievement, the day was a depressing one for Crocker. A windy rainstorm matched his mood, for it is on record that the thought of the work yet to be done discouraged him.

With the slogan, "On to Hoosac, on to the West," a Troy & Greenfield R.R. to pierce Hoosac Mountain was then proposed by Crocker. The legislature, after a brouhaha stimulated by lobbyists of the Western R.R. approved a charter for T&G in 1848. The legal specifications for the 44-mile road read, "From the terminus of the Vermont and Massachusetts Railroad at Greenfield, through the valleys of the Deerfield and Hoosac (rivers) to the state line, there to unite with a railroad leading to the city of Troy."

The Charter "pinpointed" the tunnel, "in some place between the Great Bend in the Deerfield river and the town of Florida on the east, and the base of the western side of the mountain." Two years were allotted for surveying and five for construction. The bore was to cost $2,000,000 and the right-of-way $1,500,000 more, funds to be raised from citizens and towns along the T&G's route.

On January 7, 1851, the T&G Board of Directors voted to break ground the following day, and did so on a site just west of Witt's Ledge, North Adams. Here was also the location of the first actual tunneling, since the ledge itself had to be pierced with a 324-foot "Little Tunnel," Hoosac's baby brother, so to speak.

The Yankees of Western Massachusetts had no intention of parting with their hard-earned money to back the tunnel project, witness that only $3000 in stock was sold by 1851. In desperation, T&G requested a $2,000,000 loan from the state. Answer: NO!

E. Hasket Derby recalled in later years the beginnings of T&G:

> It began in extreme poverty. . . The enterprise seemed to be upon its last feet. I recollect very well meeting with the directors one evening in a gloomy room at Earl's Coffee House, when Mr. Crocker rose from his seat and went to the mantlepiece, where two tallow candles were burning, and blew out one of them, suggesting that the road could afford but one candle in its condition at that time. . .

In spite of its slim pocketbook, the T&G bought a "Wilson's Patented Stone-Cutting Machine" built by Munn & Co. of South Boston, and by summer 1852 aimed it at Hoosac's east face. According to the calculations of Chief Engineer A. F. Edwards, this 70-to 75-ton steam-powered monster would chew a 24-foot diameter tunnel in precisely 1556 working days. Revolving iron cutters would cut a 13-inch groove for the circumference of the circle, and then, with the machine backed away from the heading, black powder would blast out the center. This $25,000 contraption ground its way 4-1/8 inches into Hoosac while the Railroad Committee of the legislature watched in fascination and delight. At this rate the tunnel would take only two years! After nosing 10 or 12 feet into the rock the machine quit, and the smiling faces fell. It never ran again and remained

The ill-starred boring machine stuck in its own hole. This engraving from a contemporary photograph appears on a "Hoosac Tunnel Series" of post cards dating from 1880. *Author's collection; see also illustrations on pages 30, 31, 32, 35, 42 and 43*

stuck in its own hole for many years.

Progress was better at North Adams. A survey had been made from the point where ground had been broken up through Williamstown out to the state line at Pownal, Vermont. Here at least something could be shown the worried stockholders. Engineer Edwards was no longer talking about finishing both road and tunnel in five years for precisely $1,948,557. He was trying to be as inconspicuous as possible after the boring-machine fiasco.

Edwards, Crocker, and backers finally realized that there would be no quick or easy way of piercing Hoosac. The only way would be the traditional hand-drill and black-powder method using one or two-man crews. In this type of drilling one man held the star-pointed drill in his hand, and with the other one hit it with a sledge, or "single-jack" hammer. When two men worked together, one held the drill and the other pounded a 20-lb. "double-jack" hammer onto it. All it took was one slip to break or crush a limb or hand. After whacking a hole two or three feet deep, the powder charge (a mixture of saltpeter, sulphur, and charcoal) was tamped into it and ignited by a candle from a powder-trail or goose-quill fuse. If the blaster wasn't a sprinter, he was maimed or dead.

From 1851 to 1854, Alvah Crocker wheeled and dealt with his legislative friends for a desperately needed state loan. After three years, he succeeded, returning in March, 1854, with a $2,000,000 loan agreement primarily resulting from the report of State Geologist Edward Hitchcock. This expert considered the bore feasible, and pressure from the many western towns finally helped loosen the legislative purse.

The $2,000,000 appropriation to T&G from Massachusetts had many strings attached. Before the state would release the first of the twenty $100,000 installments, the following work had to be done: a) the company books had to show $600,000 of stock subscribed to, b) seven miles of trackage completed and, c) 1000 feet of tunnel opened. For each additional 1000 feet of tunnel approved by the state engineer, another $100,000 would be forthcoming.

Gambling that it could raise enough backing to release the state funds, the T&G signed a $3,500,000 contract with Edward W. Serrell & Co. of Philadelphia in 1855. With this outstanding contract as proof of the promoters' good intentions, the legislature's opposition thawed slightly and it authorized the towns along the line to buy into T&G stock for up to 3% of their respective valuations. The towns were wary of the project, and with justification, for seven years of promotion found only $94,000 of stock sold, and hopes of an additional $259,283 of capital from that source soon evaporated. The situation became so desperate that T&G petitioned Massachusetts to underwrite $150,000 of stock to help meet the load requirement. The attempt failed, forcing annulment of the Serrell Contract, and work halted yet another winter.

The coming spring, the T&G went looking for a new chief engineer as well as another contractor. They found both in the person of Hermann Haupt, either the most famous—or notorious—of all the contractors who blasted at Hoosac. His *General Theory of Bridge Construction*, 1851, was well received, and is a collector's item today. Haupt was aggressive, rising quickly from surveyor to Chief Engineer for the Pennsylvania R.R. Having engineered both the Gallitzin Tunnel and Horseshoe Curve for the Pennsy, he should have been better prepared for Hoosac's challenge, but not so.

Haupt and his two partners—H. Haupt & Co.—signed a $3,880,000 contract on July 1, 1856, and brought Pennsylvania miners and Philadelphia credit to the Berkshires. His contract included all construction plus the rail to Greenfield. That would take money; and since the road didn't have any, he was also obliged to shake loose funding from the recalcitrant area towns. His bull-horn voice was persuasive to a degree, for he outstripped Alvah Crocker, selling about $175,000 of stock. Records conflict; however, it is very possible that he never saw over $125,000 for his effort. By the end of 1856, Haupt's and Crocker's combined efforts had sold $121,412 of T&G stock.

When it became evident that a railroad actually was to be built in Northwestern Massachusetts, promoters in New York State petitioned their legislature for a charter to complete the line from the Vermont border down to Troy. On November 22, 1849, a charter was granted to the Troy & Boston R.R. and on August 1, 1852, the 34-mile line was opened for business. Fellow organizers had also been busy in Vermont, creating the 7-mile Vermont Southern R.R. to connect the T&G and the T&B through Pownal, Vermont. As soon as the VS R.R. and the T&G from North Adams were completed, the T&B leased both lines and initiated through North Adams-Troy service on November 21, 1856.

The work crews inside Hoosac had little progress to brag about. Their black powder shattered the slate and mica with little trouble, but the gneiss and quartz veins could barely be cracked. The West End crews had even more problems. After penetrating the surface dirt they found a wall of limestone, holding back a water-filled mudlike compound that semi-liquified upon exposure to air. As a shovelful was removed, another flowed back to take its place.

Haupt called the substance "porridge stone," the workmen said it was like a "shovelful of eels," but the tunnel's foes had the best term for it: "demoralized rock." It made a liar out of State Geologist Hitchcock, who thought no masonry or arching necessary, by forcing the building of a brick tube six to eight layers thick for over one and a half miles of bore to hold back this muck.

Haupt, discouraged by the slow progress, fell for another boring machine—"slightly used"—and convinced the T&G that they should try it at the west heading. This machine was a 40-ton, 650 horsepower steam-driven monster constructed by the Novelty Works (appropriate name) of Philadelphia and shipped to North Adams on eight railroad cars, arriving on January 3, 1857. Designed by the same man as the first one, it was to cut an 8-foot heading with ratchets moving in concentric ovals across the entire cutting face. After fooling around for eight months, T&G first tested it on August 25, 1857, and the test was—quoting the North Adams Transcript—"not very successful." By September third, the paper reported that machine was mothballed, and the $25,000 scheme quietly died.

The West End digging was painfully slow—only 80 feet done by May 1, 1857. To speed it up, a vertical shaft was started several hundred feet east of the portal (completed September 23, 1858) allowing an additional two faces of rock to blast at.

Progress was no better in Boston than at Hoosac. Alvah Crocker had squeaked a "Tunnel Aid" bill through the statehouse, but Governor Gardner vetoed it on May 27, 1857. His "Excellency" called it "simply ridiculous" in public, and less kind adjectives in private with his friends running the Western.

Since the tunnel and state politics were becoming tightly woven together, gubernatorial candidate Nathaniel P. Banks visited the workings while stumping across the state just before the 1858 elections. With the admonition to "be sure and vote for Lincoln," in case anything should happen to him, Banks descended the shaft at the West End, viewed the heading, and talked with the miners. As the new governor, he would back the tunnel project wholeheartedly.

By now Haupt claimed in print that he was indebted for over $220,000. The T&G therefore returned to Beacon Hill after their defeat the previous year and managed—incredibly—to wheedle a $750,000 advance from the state coffers. With these funds, a $4,000,000 contract naming Haupt in charge of the entire project was negotiated in late 1858. He then let 24 sub-contracts for the grading and trackwork from Greenfield to East Portal, and work proceeded slowly and with no serious setbacks.

Haupt's major sub-contractor was a Pennsylvanian named Bernard N. Farren. By March, 1859, Haupt had contracted with Farren to also construct a 318-foot West Shaft and line 2200 feet of bore with stone masonry to block the "porridge stone."

By August, 1860, the miners had burrowed 1810 feet at the East End and another 500 feet at the West End, where timber supports were being replaced by Farren's arching. The West Shaft—7 by 14 feet—was down 170 feet.

With these results, the legislature passed another "aid" bill that fall releasing T&G from its $600,000 stock subscription requirement and establishing the post of supervising state engineer. Civil Engineer Ezra Lincoln was appointed supervisor, and he formulated plans for the systematic completion of Hoosac. Haupt worked well with Lincoln and his assistant, Stevenson. Thus when Lincoln was bed-ridden by illness in December, 1860, and Governor-elect Andrews' position on Hoosac was yet unknown, experienced politician Haupt moved quickly. In twenty-four hours he railroaded Lincoln's resignation (which he wrote!), Stevenson's appointment and a cash advance through Gov. Banks' office and Governor's Council. A masterful effort which backfired; it antagonized Andrews and fueled rising Legislative opposition. Trouble was ahead.

1861: *Massachusetts refuses Haupt's application for first $100,000 advance; work abandoned.* **1862:** *Pamphlet war between Bird and Haupt. Haupt, penniless, leaves Massachusetts and joins Union Army.* **1862:** *Massachusetts forecloses on T&G mortgage.* **1862:** *Massachusetts Legislature forms Tunnel Commission.* **1863:** *Favorable Commission report; heated debate follows.* **1863:** *Tunneling starts under state management; Central Shaft begun.* **1866:** *First American pneumatic drill used at Hoosac.* **1866:** *First nitroglycerin tested as replacement explosive for black powder.* **1866:** *Massachusetts aroused by two more of Bird's pamphlets.* **1866:** *Start of brick arching at West End.* **1867:** *Fire at Central Shaft leaves thirteen dead and Legislature in an uproar.*

CHAPTER  THE STRUGGLE: 1861-1867

The progress on Hoosac and T&G was met by the hearty disapproval of the Western R.R. management. Their monopoly was threatened at a time when their political leverage was weak, since Governor Gardner had been succeeded by N.P. Banks. In desperation, President Chester Chapin called on Frank W. Bird of Walpole, Mass., to aid them.

Bird, a wealthy paper manufacturer and political sleight-of-hand artist, wrote "with a pen dipped in sulphuric acid, crossed his 'T's' with arsenic, dotted his 'I's' with Paris Green, and blotted the paper with rat poison." He was sympathetic with the tunnel's opponents led by Daniel Harris, and with Western RR backing he quickly raised merry hell in the Statehouse about the "Great Bore."

By May, 1861, Bird had managed to deal Replacement State Engineer Stevenson out of his post and have a crony—William S. Whitwell—appointed successor. Whitwell was to slow Haupt down in any way possible while Bird attempted to kill the project's state funds.

Haupt and the T&G felt the effects of this political football, but were powerless to retaliate. Disaster struck when Haupt finally completed enough work to qualify for the first $100,000 loan. The work had been approved by Stevenson and the payment draft lay on Governor Banks's desk. In a classic blunder, Banks became so involved in his farewell speech that he forgot to sign the draft. Incoming Governor Andrews—an initial friend of Bird—refused to honor it without State Engineer Whitwell's signature. Whitwell refused. With creditors hounding him, Haupt abandoned the tunnel work in midsummer 1861 to become Brigadier General in charge of Construction and Transportation for U.S. Military Railroads.

With this knock-out blow, the project seemed doomed. A forlorn "Transcript" editor noted there were entirely too many "Bird" tracks in the mud around the "Great Enterprise" as the sly foe continued to gather evidence to support his point of view. Using rocks he gathered from one of those trips, Bird would stop a Congressman, click the two stones under his nose and query, "Would you build a railroad into a closet?"

By early spring, 1862, Bird published—at the taxpayer's expense—*The Road to Ruin;* a pamphlet describing the ills of the T&G in general and Hoosac Tunnel in particular. Haupt's abortive use of steam drill was just one of its many targets. This gem was immediately followed by *Fact vs. Illusion*, in case anyone was left in doubt.

Haupt, shocked by the vehemence of this absentee attack, quickly fought back. *The Road to Ruin, or The Decline and Fall of the Hoosac Tunnel* was immediately followed by Haupt's *The Rise and Progress of the Hoosac Tunnel.* To quote its first paragraph:

> In a recent interview with a learned judge in Boston, your pamphlet formed the subject of conversation, and. . .that gentleman expressed the opinion that Frank Bird was an 'honest man.' I was greatly surprised at such a declaration. . .

To combat the charges of incompetency in *Fact vs. Illusion*, Haupt penned *H. Haupt's Memorial.* He presented it to the legislature in desperation since he needed the state funds to cover part of the over $1,225,000 his Philadelphia creditors had advanced him. Charges and countercharges from Haupt and Harris flew under the "Golden Dome" and Haupt left Massachusetts penniless. He wouldn't get a settlement of 8¢ on the dollar until 1884.

With Haupt's ouster, the legislature started reviewing the progress of the T&G and its "Great Bore." One item was the Green River bridge west of Greenfield. The 273-foot bridge built and designed by Haupt, 1) was on a curve, 2) lacked structural stability, 3) had the main suspension rods anchored in pine rather than hardwood, 4) was top-heavy, and 5) promptly fell down under its first train, killing one man and injuring two more. Small wonder *Fact vs. Illusion* recommended that Haupt be the first engineer over any bridge he built for the Union Army.

Haupt's general construction record was poor. His trackage from the bridge to West Deerfield—7 miles—was on tempo-

Fragment of old map shows T&G line west from point J to West Deerfield. Arrow marks site of the Haupt bridge that collapsed. Dotted lines are proposed routes for the Boston, Hoosac Tunnel & Western RR. Points I, D, E indicate present-day Boston & Maine RR location. *Author's collection*

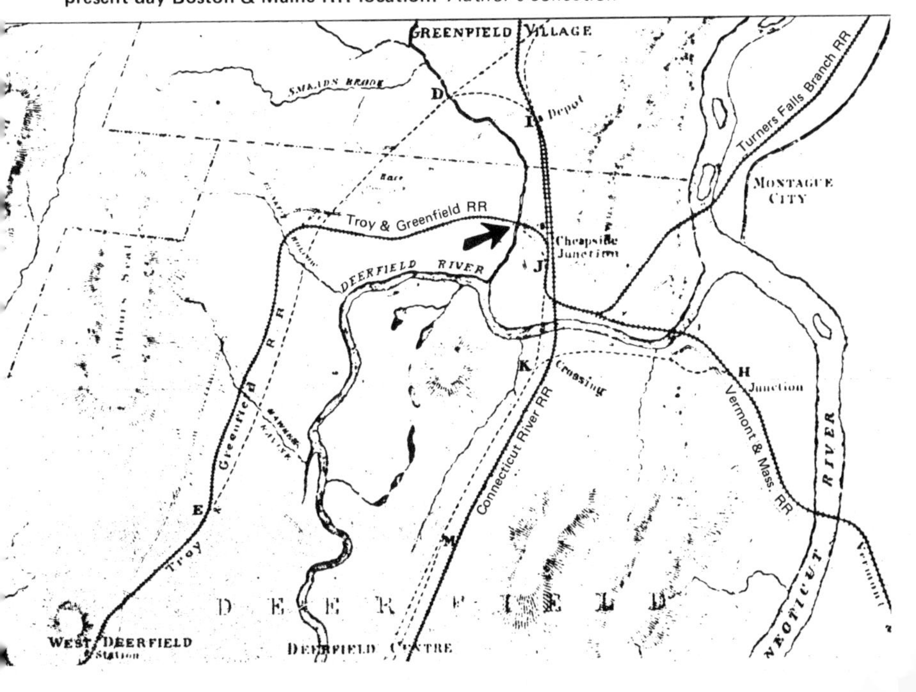

rary trestle work over the depressions, rather than on fill, and embankments were on a 3 to 1 slope, rather than the more stable 2 to 1. No fill going into the Deerfield River was even rip-rapped. His five years of boring left 2400 feet blasted at East End, 610 feet dug from the West End, and the West Shaft down to grade—318 feet—and continued 561 feet west. The west heading was a 16-foot stone arch for about 100 feet, and the other 500 feet was just supported on timbers; much too small for double tracks.

The most damaging evidence against Haupt was that of 7581 shares of T&G stock, he owned 5987, leaving various people and towns controlling a block of 1679 shares. Alvah Crocker was T&G's second largest stockholder with *15*!

Examination of the records today indicates that Haupt was simply overwhelmed by the project; not that he attempted to defraud Massachusetts. His overwhelming stock control resulted from a willingness to take payment in script—the T&G had no cash—just to keep the project going. His intentions were getting as much rock removed and rail down as possible. Improvements could come after getting the state loan. What discredited Haupt was his inadequate bookkeeping, for he could not disprove Bird's charges.

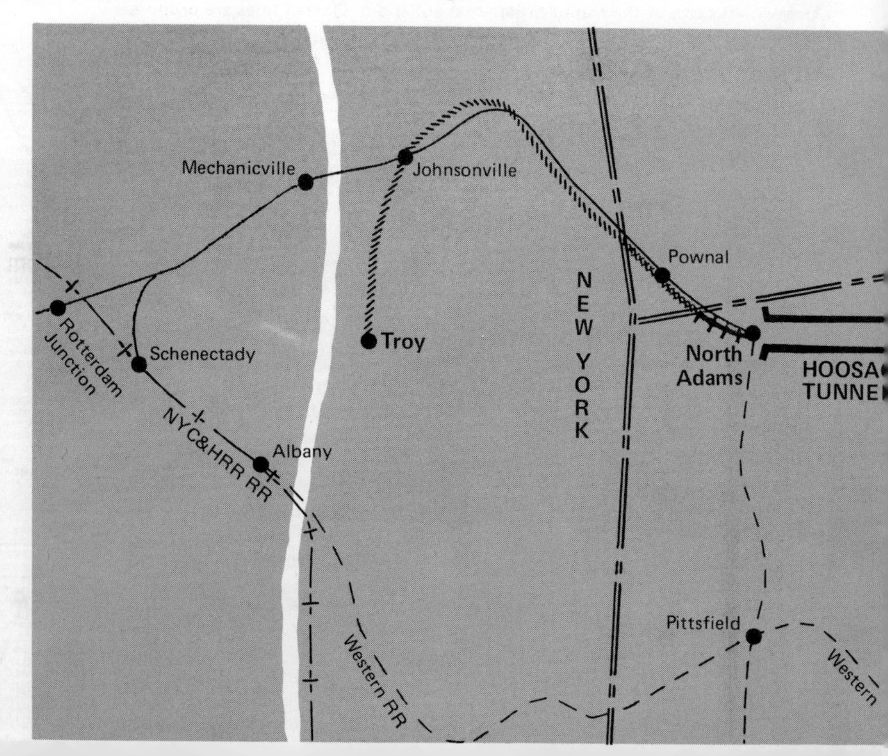

In the midst of this boondoggle, the thrice-mortgaged T&G—May 28, 1855; August 18, 1855; June 6, 1860—collapsed. It defaulted on its mortgage payments August 18, 1862, and the Commonwealth took control September fourth.

With Massachusetts in the tunnel business, the legislature formed a Commission to study the project and report the following spring. Hoosac lay dormant for another long winter.

On March 19, 1863, Commissioners John W. Brooks, Samuel N. Felton, and Alexander Holmes submitted their report "establishing the feasibility of the grand enterprise of tunneling the Hoosac Mountain." Their major recommendations, resulting from Consulting Engineer Charles Storrow's trip studying European tunneling methods, were drilling by compressed air and replacing black powder with nitroglycerin. Much of Haupt's work would have to be redone, for he had not followed Edwards's survey closely, and two separate tunnels were taking shape. A new West Portal was necessary, and the gradients would have to be corrected. Both headings would have to be widened for double tracking, and the West End needed brick lining.

Map drawn specially for *A Pinprick of Light* by Richard W. Symmes

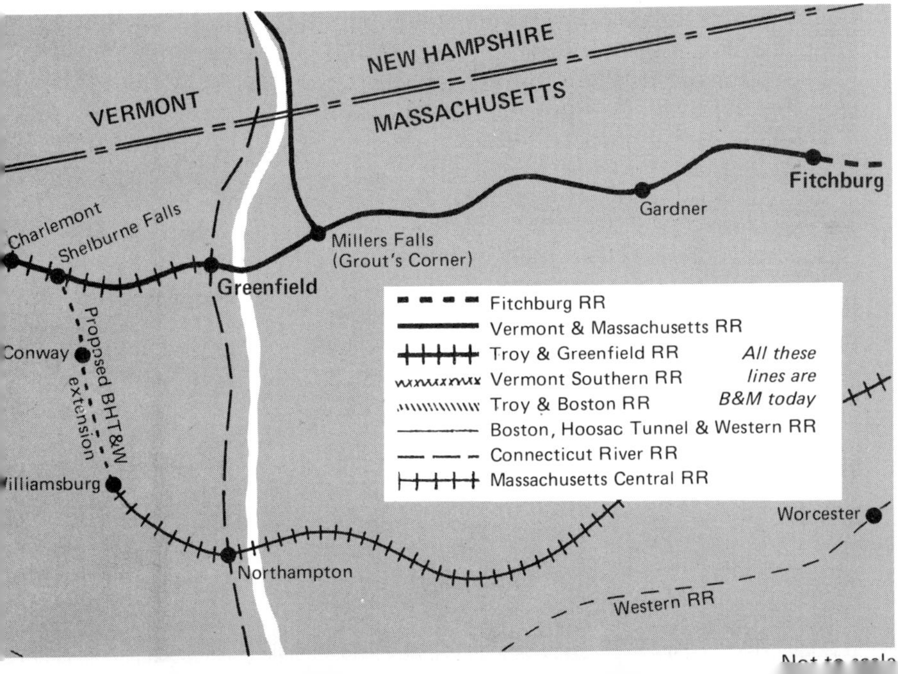

By this time Massachusetts had spent $954,693—less interest—and the consulting engineers couldn't agree on a projected final cost. Estimates ran from $2,800,000 to $3,400,000 for Hoosac alone and up to a grand total of $5,750,000 for T&G and its bore. The most accurate estimate by far belonged to Frank Bird: $10,000,000+.

The Commissioners were authorized to begin immediate construction with payments of "just claims" coming from the Governor's Council. On July 1, 1863, Chief Commissioner Brooks hired Charlestown Civil Engineer Thomas Doane for chief engineer and three weeks later got the first $15,000 to start work again.

During the two year political hassle, Haupt's charges of collusion between Bird and State Engineer Whitwell were forgotten. Tom Doane soon remembered, for Whitwell immediately demanded the construction of a "central shaft" between Hoosac's peaks to "speed up the digging." This 27 by 15-foot shaft would drop from the cradle where the Cold River forms down over 1000 feet to connect with the exact center of the bore. The unconvinced chief engineer first had to relocate the shaft some 600 feet west of center, for as Whitwell—and Bird—knew, the exact center of Hoosac is in the middle of the Cold River! Hoist buildings were erected, and groups of ten to fifteen miners worked around the clock sinking central shaft to grade.

State Engineer Whitwell also advocated Consulting Engineer James Laurie's recommendation that the T&G main line go straight from East Deerfield across the meadows and up the river valley. This route had its merits: 150 feet less grade, 489° less curvature, and a mile shorter. However, the Greenfield loop was in existence, and with T&G's bankruptcy, it was but a delaying tactic.

Late 1863 to 1866 was a quiet period in Hoosac's construction. The Civil War preoccupied Massachusetts; so as bills arose, the Commissioners quietly paid them. By December, 1864, only 1145 feet of heading had been opened under the Commissioners' control. Costs were skyrocketing, and by December, 1865, the state had spent $2,324,072 in cash and invested another $494,517 in machinery.

The long, costly project had become the boondoggle of Massachusetts. "The logroll became a fine art on Beacon Hill during the wheeling and dealing over it," and Oliver Wendell Holmes even penned this poem:

> When publishers no longer steal
> And pay for what they stole before,
> When the first locomotive wheels
> Roll through the Hoosac Tunnel bore,
>
> Till then let Cummings blaze away,
> And Miller's saints blow up the globe;
> But when you see that blessed day,
> Then order your Ascension robes.

Distrustful of ascension robes, Doane decided to put Storrow's recommendations into practice. A rock crib dam thrown across the Deerfield River about a mile above East End channelled a 16-foot head of water into a large stone mill holding air compressors he designed. June 14, 1866, the first American pneumatic jackhammer bit into Hoosac more readily than any star-pointed drill.

These drills' development dated from 1838 when Issac Singer (of sewing machine fame) built two semipractical steam drills. In 1859 a steam-powered drill was tried at Hoosac, but was too heavy, and its exhaust nearly suffocated the men. Haupt started to modify it for air operation, but he never got past the planning stage. By 1861, Germann Sommeiller, chief engineer at the Alps' Mt. Cenis tunnel had a monstrous but practical drill operating. Charles Burleigh of Fitchburg redesigned these drills, and soon his Burleigh Rock Drill Co. was producing a lighter and more efficient drill with Doane's air compressor to power it. The East End drills were mounted on movable carriages and connected by rubber hose and iron pipe to the water-powered compressors. Doane said, "Mr. Burleigh's drills have many 'weakest points,' and they break down often," but improved progress immediately.

Doane resurveyed the tunnel line and replaced Edwards's grooved iron markers with six stone "lining towers" during 1866, one on each peak of Hoosac, one at each portal, and

BERNARD N. FARREN *Courtesy of Mrs. Percy Bugbee*

Below: **What the well-dressed surveyor wore, circa 1868. This group is posed near the chief engineer's office located just above the canal to the power house at East End.** *Henry Ward photo, courtesy of New England Power Company*

one on Rowe's Neck to the east and Notch Mountain to the west. The crews cast a line from their 12-foot square tower to the 25-foot-high red and white iron pole at the peak of the sloped roof on the next tower in line, or into the black portals where it was carried forward on plumb bobs driven into wooden plugs in the cavern's roof. The line was extended by sighting down all the plumb bobs toward a tiny light glowing in the blackness. This light was moved and sighted, moved and resighted, until the line was straight as an arrow. Then a miner's lamp flashed once, and the drillers went back to work.

The West End's "demoralized rock" flooded Haupt's tunnel, and only Farren's stone arch served to drain the surrounding earth. After the muck collapsed countless timber supports, Doane rehired B.N. Farren in 1866, this time to build 500 feet of six-to-eight-layer thick brick tubing. Completed it stood 24 feet high and 26 feet wide and rested in a hand-gouged hopper 550 feet long, 300 feet wide, and 75 feet deep. This tube was continued for 883 feet with dirt covering the bottom and trackage laid upon it. 7573 feet of Hoosac in all were bricked—the rest a simple brick arch—and resulted in a brick yard on the flat between West Portal and West Shaft firing over 20 million bricks in six years of operation.

The West Shaft was where all eastward blasting was done. Even there the water flowed from the heading at 600 to 1000 gallons per minute while steam hoists and pumps fought the flow. To investigate the rock ahead, Doane ordered a series of four small shafts, or wells, dug to grade. One of them, #4, was brick-lined, and about ten feet square, often referred to as the Brick or Baby Shaft. In addition to the wells, the Supplementary Shaft was sunk for pumping lines while the spoil came up in hoists and buckets in West Shaft. All these shafts were filled in when Hoosac was completed.

By now the boys on Beacon Hill could once again settle down to their traditional politics. Frank Bird dashed off another mudslinger—*The Hoosac Tunnel, Our Financial Maelstrom*, featuring a special broadside at the Deerfield dam and powerhouse. Among other statements, Bird "guessed if the Deerfield were allowed to run by day it would all evaporate before it reached Shelburne Falls." The river contradicted him by powering the eastern heading, but steam power was required at the Central and West Shafts.

Alvah Crocker replaced John Brooks as Chief Commissioner and disarmed Bird's attack by arranging the $30,000 annual lease of the T&G and V&M to the Fitchburg R.R.; Massachusetts was thus getting some return on its ever-increasing tunnel investment.

By 1867 the crews were averaging 80 to 100 feet monthly, and more Burleigh drills were on the way. July 31st found trackage completed west to Bardwells Ferry, and the rebuilt Green River Bridge opened. The Central Shaft was over 520-feet down, and by August 23rd, Farren had completed 450 feet of brick arching and had contracted for 500 feet more. A new 50-horsepower boiler was running a 1200-gallon-per-hour pump at West Shaft and a contract for the West Shaft heading with Dull, Gowan & Co. was signed.

Between the mounting death toll and Massachusetts' lackadaisical prosecution of the work, Tom Doane finally quit in disgust. After publically venting his anger and frustration on Alvah Crocker, he went on to become the Burlington R.R.'s chief engineer. C.P. Granger replaced him in August, 1867.

The rockmen kept hammering away at the quartz-lined rock. Huge timber gates half-protected them from flying debris as the powdermen did their job. The din was overwhelming with shouted commands, crashing explosives, chattering jackhammers and hissing air pipes. Everyone was always soaked to the skin.

The ingenious engineering of Central Shaft was left to a Dutchman named Carl O. Weiderkinch. His job was to insure that Central Shaft would reach grade squarely in the middle of the tunnel centerline and that the headings would be on target with those from the portals. Weiderkinch constructed two masonry piers along the tunnel centerline each 25 feet from the shaft's lip. Capping each pier was an adjustable iron block with a thin vertical slit for sighting in the lining towers. When the alignment was perfect, two piano wires 1/20th of

West Portal, 1866. Beginning the brick arching. Note the surveyor and the mule carts (the mules were stabled in Haupt's abandoned stone bore).

Below: East Portal, circa 1867. The "lining (survey) house" of stone with its red and white pole is directly in front of the portal and the abandoned first bore (at left center). *Both photos by Henry Ward, courtesy of New England Power Co.*

West Shaft buildings

an inch apart were stretched between the slits. Between these two wires hung two 15-lb. plumb bobs 25 feet apart and perfectly in line with the towers. Both sets of plumb bobs were hung in 8-inch square wooden casings terminating in containers of light oil to minimize vibration. Even then the lines oscillated an intolerable 1/100th of an inch and had to be corrected by averaging several hundred observations when grade was reached. Engineer Weiderkinch's patience and accuracy paid off; for when his lines were carried to West Portal from Central shaft the error was 9/16" and an error of only 5/16" when projected to East Portal.

Central Shaft was a killer. Workmen rode to the heading in a swaying iron bucket with no escape possible from anything that might plummet down upon them. Both men and rock went to the surface this way; and if the hemp rope failed, God help everybody! Once a newly-sharpened drill rolled into the abyss and impaled a rockman from head-to-toe over 300 feet below. Later some unfortunate soul fell down the entire 1028-foot chimney after which he was "rolled up like a side of leather" and taken to the funeral parlor.

Hoosac's worst disaster occurred on October 17, 1867. The Shaft House basement held an abandoned "gasometer" that had been tried for illuminating purposes. About 1 p.m. a blast was fired and 13 miners descended to shovel up the

Central Shaft buildings erected after the fire in 1867

Above: Miners coming to surface in elevator, from "Hoosac Tunnel Series" of post cards. Post card cross section art of Central Shaft, *right,* is dated "at the time of the fire," 1867

Miners working at 583-foot depth

debris 583 feet below. For some reason, a worker opened the door to the hoist-house basement and the naptha fumes exploded, soon sending over 300 drills and the flaming building down onto the unsuspecting men. By 3 a.m., a miner named Mallery was lowered into the pit with only a rope around his waist and was hauled up unconscious. Having seen only burnt timbers and water, he gasped out "no hope" upon being revived.

Central Shaft filled with water, and several bodies surfaced. However, the remaining bodies would not be reached until October 19, 1868, when it was discovered that not all the men had been crushed instantly, but that some had built a raft, only to be asphyxiated by the fumes.

In addition to the deaths, between $40,000 and $60,000 in buildings and material was burned. Work stopped at Central Shaft amid cries to abandon the "bloody pit" once and for all. Faster and safer progress would have to be made if the Hoosac Tunnel was to be finished.

Post card series engraving shows details of first work on brick arch rather more clearly than the picture on page 29. Missing, however, are the four ladies posed at the top of the arch in the photo.

1867: *Mowbray starts producing tri-nitroglycerin.* **1868:** *Bird publishes two more pamphlets; rebuttal pamphlet starts another debate in the Legislature. Result: Bird disappears and state funds cut off for Hoosac.* **1868:** *Massachusetts signs a $5,000,000 contract with F. & W. Shanly Co. to complete tunnel by March 1, 1874.* **1869–1872:** *Period of steady work under Shanly contract.* **1872:** *Blasting from Central Shaft to West End stopped due to potential flooding.* **1872:** *Brick arching completed.* **1872:** *Central Shaft to East End opened December 12. West End blasting immediately resumed.* **1873:** *On November 27 (Thanksgiving Day), Hoosac Tunnel completed.*

CHAPTER  TRIUMPH: 1867–1873

Blasting had to be speeded up. Black powder simply could not shatter the quartz veins in Hoosac. Thus an 1866 advertisement in *Scientific American* caught Tom Doane's eye. Perhaps George Mowbray's explosive could help. Doane invited him to come to North Adams.

Mowbray, an English chemist specializing in explosives, had come to America in 1854, hoping to become wealthy in the Titusville, Pennsylvania oil business. He knew of Sobrero's experiments with nitric acid, glycerin, and sulphuric acid culminating in the 1846 discovery of a highly unstable "soup" he called "nitroglycerin." Mowbray had added some refinements to the process, and was prepared to concoct that era's most powerful explosive: "tri-nitroglycerin."

Nitroglycerin was considered so unstable that it was banned in Europe and no railroad in America would carry it. Its reputation was deserved. America's first experience came in 1865 when a man in the lobby of New York's Wyoming Hotel noticed a reddish vapor coming from a small package. He remarked about it, and the desk clerk heaved the package containing "Glonoin Oil"—a form of nitroglycerin—into the gutter. The explosion broke every window within 100 yards, cracked the pavement, and flattened several people. The following year a similar package destroyed the Wells Fargo office in San Francisco, leaving eight dead and $2,500,000 damage.

Doane was doubtful, for nitroglycerin compounded by "Colonel" Tal P. Shaffner and tried at West Shaft during July and August, 1866 performed no better than powder. However, it was yellowish and clouded with impurities when compared with Mowbray's crystal-clear liquid.

Arriving in North Adams on October 29, 1867, Mowbray quickly started building a two-story factory about 100 feet south of West Shaft and a small home 20 feet further away. By December 31st, the completed factory was in operation.

The "Acid House," as it was nicknamed, had stove heating until the newfangled steam heating could be installed. This 150-foot building housed the soapstone troughs where the acid mixture was prepared. This mixture was carried into the 100-foot-long "Converting Room" where 116 stone pitchers sat in nine ice-water-filled troughs. Pure glycerin from glass jars shelved above the troughs dripped slowly into the acid-filled jars while the ensuing fumes were blown away. If the fumes built up because of a fan failure, explosion would result.

When the fumes cleared, indicating the completed chemical action, the jars were carefully emptied into a water tank held at 70°F. The nitroglycerin slowly settled to the bottom of the six-foot tank and then moved into a butter churnlike cistern where an air current stirred and purified the explosive. The water was discharged through three barrel traps to capture any remaining nitroglycerin and then into a rock pile. A man now carried the two copper pails of liquid nitroglycerin—very carefully—on a shoulder yoke to the magazine 300 feet away. There it was put into one of twenty earthenware crocks submerged in 70° water constantly warmed by a steam pipe. Three days later the impurities were skimmed off, the nitroglycerin packed in 56-lb. waxed tins, and frozen by ice and salt. There the nitroglycerin stayed

Mowbray's "Acid House." "Dangerous!!" warns the sign, but the fellow lounging in the road seems not to be very concerned about it. *Henry Ward photo; Ruth B. Browne collection, North Adams Public Library*

Below: **The converting room in the tri-nitroglycerin factory.**

until going to the heading wrapped in rubber tubes, sponges, and covered with a layer of ice. Running warm water through the rubber hoses liquified the explosive, making it ready for use.

Mowbray's first winter in North Adams taught him a lesson about his "soup," for originally he thought it was most stable at 90°F. Thus one wintry day, C.P. Granger started up Hoosac with a load of carefully warmed explosive-filled cartridges destined for the ice-blocked Deerfield Dam. Suddenly the sleigh skidded over a snowbank and Granger jumped, convinced Eternity was only moments away. Feeling no explosion, he gingerly crawled out of the snowbank, brushed himself off, and found that the nitroglycerin was frozen solid. Granger then continued to East End with the cartridges stuffed in snow between his feet where he found that even exploders would not detonate the rock-hard nitroglycerin. From then on, the liquid was always moved or stored frozen under ice.

The blasters quickly complained that tri-nitroglycerin was no better than black powder. An irate Professor Mowbray found that the lazy blasters were not drilling the deeper holes—42 inches vs. 30 inches—he had carefully specified. Following directions, they soon found the explosive much more powerful and safer than powder; a scheduled comparison of the two was never done, for the men refused to go back to the "dangerous" powder!

A better means of detonation than a powder trail was needed with nitroglycerin. Various experiments were tried and the best method seemed to be the friction machine, which ignited exploders by static electricity. "Colonel" Shaffner had brought a friction machine that would ignite five charges simultaneously. In 1868, an "Austrian" machine was imported that would fire twenty to thirty charges over 500 feet of lead wire. Still unsatisfied, Mowbray upgraded this machine to fire any number of charges desired, and from the timekeeper's office at West Shaft, nearly 12,000 feet from the final West Heading.

Brothers in North Adams developed the first "safe" nitroglycerin exploders and insulated lead wires. The exploder consisted of two insulated wires terminating in a hollowed-out chamber in a wooden plug with a pasteboard chunk covered with fulminate of copper placed between the wires. The Browne brothers also insulated copper wire with gutta-percha. Blocks of gutta-percha were cut, rasped, rolled, washed, and then forced around the wire at "95-tons" pressure. Indicating the electrical knowledge of the period, the lead wires were coated with twice the insulation of the return ones! The Browne brothers were so successful that their wire and exploder factory ran for several years after Hoosac was completed.

Contractors Dull, Gowan & Co. at West Shaft went bankrupt attempting to meet their contract's 100-foot-per-month progress stipulation and quit. This precipitated a resignation from the Tunnel Commission and more political fireworks in Boston.

By June, 1868, the state miners were averaging 150-feet per month, but still 16,800 feet remained. "B.N." Farren's crews had opened the trackage from Greenfield to Shelburne Falls on January 1st, and completed it to East Portal on August 23, 1868.

Travelers could now go between Boston and Troy by rail, except at Hoosac. There teams of six white horses lifted the traveler up the road from Hoosac Tunnel Village or North Adams and over the twin peaks. In the steep downgrade, the Concord stagecoaches hauled logs behind them to keep under control—a magnificent sight.

The tunnel opposition published a pair of Bird's now well-known pamphlets: *The Last Agony of the Great Bore*, and *The Modern Minotaur*. Among other complaints, Bird bewailed that the interest on the invested state funds was greater than the total state education budget.

The opposition had gone too far, and an author calling himself "Theseus" attacked Bird's efforts with a barrage of facts, figures, and a long memory. He reminded the Massachusetts Yankees in the midst of Reconstruction that gubernatorial Candidate Bird once had Southern sympathies. It seems Bird had been chosen in 1862 by Governor Andrews for his Council, but refused to swear allegiance to the Constitution. Disgusted political opponent Tappen Wentworth informed Bird, "Take the oath or be thrown out of office!" "The office predominated over principle," Wentworth observed, and Bird swore the oath.

"Theseus" described Bird's latest pamphlet:

> The gory locks of the Hoosac Tunnel have given Frank Bird, the Walpole dyspeptic, another turn, and he pours out 53 pages of pecksniffian lamentations and abuse, drawn from the never-failing fountains of spleen and remorse which constantly engulfs his guilty mind. . . [Bird's briefs], filled with the gentle breathings of the sweet spirit of the Walpolean paper maker and pamphleteer, were yearly broadcasts on the legislative ocean. He now. . .testifies. . .in. . .print at his own, or some other person's expense, the years he laboured in company with his Sancho Panza from the Connecticut River to vilify, blacken, and destroy the reputation of Haupt & Co. . .With the malice of a fiend, he haunted the State House from the Coal Hole to the Dome, prostituting every faculty to the accomplishment of his scheme, which was the ruin of Haupt & Co., and stopping the work on the road and tunnel.
>
> He succeeded, then, as now, in getting committees of investigation appointed. Every Committee returned unanimous reports in favor of Haupt & Co. . . .and opposed by none save Don and Sancho and their 7-by-9 organ, the *Springfield Republican* . . .Bird and Co. saw two ways of killing the tunnel. . .stop it (or) increase its dimensions (and) sink a central shaft. . .

With Massachusetts politics again in an uproar over Hoosac, Frank Bird finally disappeared after being roundly defeated for governor, but not before state funds were cut off on October 16, 1868. In one final attempt to keep Hoosac alive, ringleaders Crocker, Felton, and Wentworth persuaded the legislature to put the entire job up for contract. Dillon & Co. of Chicago and F. & W. Shanly & Co.

of Montreal bid for the contract to complete the bore and lay one track by March 1, 1874. On January 7, 1869 Governor Bullock announced:

> The contract for the construction of the Hoosac Tunnel was executed on the twenty-fourth of December. . .with Walter Shanly of Montreal and Francis Shanly of Toronto, Canada, for the sum of $4,598,268. . .

The original contract stipulated that the contractor deposit $500,000 security with the state—an exorbitant fee which Governor Bullock thought certainly would prohibit any bids for the job. However, Shanly's offered to do $500,000 worth of work before Massachusetts paid them anything, outfoxing Bullock at his own game. The contract also had such punitive clauses dealing with failure to meet the minimum monthly heading advances, however, that both the Governor's Council and Shanly's immediately rejected it. Consulting Engineer Latrobe, who had drafted the contract, left in a huff, State Engineer Frost was ignored, and Tunnel Commissioner Wentworth drew up a final contract acceptable to all.

By March 29th, the "Shanly"—as the workmen called the brothers—had men working at East End and a week later at Central Shaft. French-Canadians, Irish, Cornish, Italian, and Yankees were hammering away hooking up drainage pumps and building some twenty wooden platforms for Central Shaft. A half-ton of candles for illumination was ordered and Walter Shanly ordered the Shaft House rebuilt with self-closing iron hatches in a fireproof floor to allay fears of another fire. By April 19th, seventy-five men and six steam boilers were working at West End with four new compressors and 150 more men coming.

B.N. Farren completed his brick lining at West End on February 8, 1869. His crews were not affected by the stop-work order from the state and had been the only men working all winter. Shanly's subcontracted their brick work to Hawkens & Holebrook Co.

Mowbray picked up one of the mule teams from the state and started using it to cart nitroglycerin to Central Shaft and

"THE SHANLY." Walter and Francis, the principals of F. & W. Shanly & Co, who punched through "the Great Bore" ahead of their contractual deadline. *North Adams Public Library*

At right, opposite page: **Actual shot of Burleigh drills in action in Hoosac. The fog was created by the photographer's magnesium flash powder exploding.** *Henry Ward photo; Ruth B. Browne collection, North Adams Public Library*

East End. The townspeople feared him and his "soup" alike, giving both a wide berth and voicing the opinion that anybody crazy enough to work with them deserved what they got. Thus one Sunday when he drove to the icehouse to get ice to pack his load of nitroglycerin in for a trip to East End, Mowbray was refused. The righteous merchant didn't do business on "The Lord's Day." Exasperated, Mowbray unhitched his mules and threatened to return them to the stable, leaving a wagonload of explosive in front of the icehouse. The ice appeared.

Several days of rain flooded the brook over the West End on October 4th. The endangered workmen sent a horse charging up to West Shaft and came pounding out on its heels. The hoists ran up and down once a minute until all 100 men were out safely. Hoosac soon filled to within 18 inches

of its roof, and much of Farren's brickwork had to be replaced at Shanly's expense. The same storm also washed out the line from Greenfield to East End.

Hoosac Tunnel Village was growing fast with Shanly's providing homes, a schoolhouse, and a meetinghouse. In this era before accident insurance, Walter and Francis Shanly paid medical or funeral expenses for the injured workers and helped support many a deceased miner's family. In so doing, these compassionate brothers earned the great respect and loyalty of their employees and associates.

During May, 1870, tri-nitroglycerin was pitted against guncotton, which is nitroglycerin-soaked cotton, and Dualin, a mixture of nitroglycerin and sawdust. The guncotton was weak, while the Dualin's fumes were more potent than its explosive potential—overcoming both workers and the doctors sent down to treat them. The tests were quickly abandoned, and soon Mowbray was producing 250 lbs. of nitroglycerin daily.

On Independence Day, 1870, the first train since the flood reached Hoosac Tunnel Village, and on August 13th Central Shaft reached grade. The shaft's precarious bucket arrangement was removed, and one of the first Otis Industrial Elevators was installed.

At East End, the blasting proceeded at grade level with a second blasting area about 600 feet behind the heading that enlarged or "sloped out" the roof to full size. At West End the process was reversed, with the heading being cut at roof level, and the tunnel floor or "bench work" cut to grade behind it.

The rock and water removal presented a problem. Walter Shanly finally ordered a ten-ton locomotive late in 1870 for East Portal, to replace its mule teams. A double-lift was installed at West Shaft to speed rock removal, for with brick lining still being done, little spoil could go out via West Portal. Central Shaft had eight 214 gallon-per-minute pumps in operation, but still the water increased. Some single veins between West Heading and Central Shaft spurted over 3000 gallons every hour, and men were nearly drowned as

Tunnel enlargement, West End work

Tunnel heading, showing drill carriages

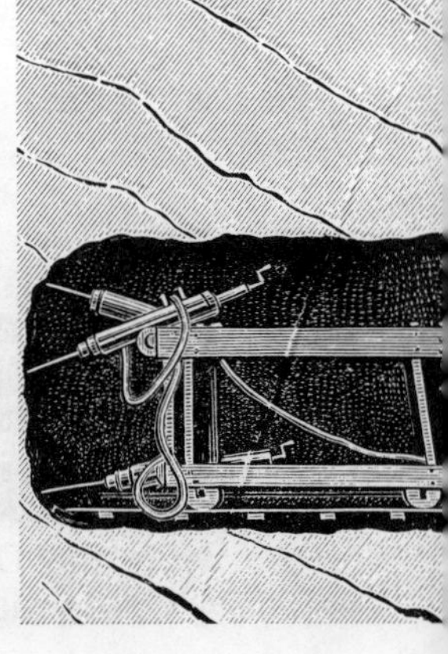

nitroglycerin opened more veins unexpectedly.

The danger of Central Shaft flooding forced Walter Shanly—business had called Francis Shanly to Toronto—to petition for the abandonment of the western Central Shaft heading until the East End heading was met. State Engineer Frost forbade the move, even with doubled rates of progress at the other headings, and forced Shanly to purchase more pumps costing $217,000.

Even the new pumps were overwhelmed, and finally in May, 1872, Shanly abandoned the heading. With Frost screaming in protest, he hurried to Boston and placed the problem in the lap of the Governor's Council. They realized the pigheadedness of Frost's position and approved Walter Shanly's decision—but wouldn't pay for the unnecessary pumps!

Momentum peaked that year with over 900 men working in three 8-hour shifts. East End alone had 350 workers, a locomotive, 40 dump cars, and a machinist-blacksmith shop located in the abandoned bore. Wages for that year—in 1870 *dollars*—totaled over $500,000. On December 12th, a final charge of nitroglycerin opened the wall between East End and Central Shaft, making a pinprick of light more than halfway through Hoosac.

Roof enlargement, East End work

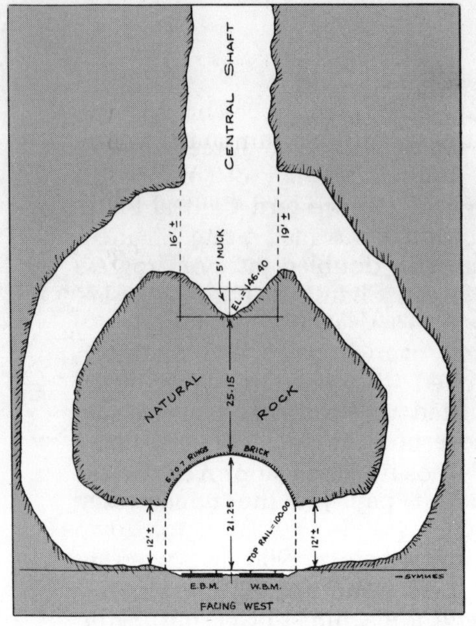

Hoosac Tunnel cross section at Central Shaft, facing west. Though the diagram doesn't suggest it, the twin ventilating shafts are offset where they join the tunnel: the left passage 33 feet eastward of the main shaft, the other 35 feet to the west. *Drawing by Richard W. Symmes, courtesy of Boston & Maine Railroad Historical Society, Inc.*

Security precautions were not like today's, and two local boys—H. Frank Blanchard and Dallas A. Dean—donned miners' caps and rode down Central Shaft like workmen. They promptly scrambled through the broken wall and became the first men through that half of the bore.

With over half of Hoosac complete, the pumps no longer had to lift all the water up Central Shaft, but could just pump it over the crest and let it flow out East Portal. The brick arching was completed on July 5, 1872, and the full crews now tackled the watery rock and poured spoil out both portals.

Sub-contractor C. McClallan & Sons were building the 1.5-mile rail line from the North Adams Depot to West Portal. Using one of the area's first steam shovels, they quickly demolished the remnant of Haupt's tunnel. As the stone arch collapsed, everyone suddenly realized that the West End crews would be idle until the mess was cleared away. Rocks started going up in the buckets at West Shaft again as Messrs. McClallan were called to Walter Shanly's office for an enlightening conversation.

On November 26, 1873, the blasters reported they only

had 16 feet of heading left. An immediate celebration was called at Rowe Town Meeting Hall. A triumphant Walter Shanly announced to his partner as he danced the evening away, "Miss Clara, if we have good luck, we'll shake hands through the tunnel tomorrow."

At 6 a.m. the following morning—Thanksgiving Day, 1873, appropriately—a two-inch speaking tube was cut through the headings. Later, Professor Mowbray arrived with 155 lbs. of his cartridges and placed them on both sides of the heading. The blaster was 370 feet east of the rock wall, and all exploders were wired, the West End's through the speaking tube, to it. About 3 p.m. the timber gates were closed across the tunnel some 300 feet from the barrier. As the official party including Walter Shanly, Colonel C. P. Granger, and Tunnel Commission Chairman Robert Johnston came down Central Shaft elevator, an expectant hush settled over the crowd.

Inside connected 4-4-0 wood burner used as construction locomotive. *William Fletcher collection*

Walter Shanly plunged the blaster handle home, and as the explosion died away to both East and West, it was met by a tremendous cheer from the 500 onlookers buried under carts and behind equipment. Every light was out, it was bitter cold, and a drill carriage had been smashed by a flying boulder, but it didn't matter: Hoosac Tunnel was a reality!

At 3:15 p.m., Shanly led the march to the 5-by-5½ foot crater, but to the workers' disgust, turned aside to let Commissioner Johnston pass first. Shanly, Granger, the blasters, and then the general public passed through. Reporter Manley M. Gillam was the first man to pass completely through Hoosac, for he came in East End and rushed out West Portal to reach the North Adams telegraph station with his story for the Boston papers.

Wheeler's Brass Band serenaded the people as they streamed from West Portal to the Arnold House for a reception and then a grand ball that evening at the Opera House. Ironically, the man most responsible for Hoosac's success was absent. Alvah Crocker had been elected to Congress in 1872 and was in Washington, D.C. Although he would walk the bore, he was not destined to ride through it, passing away December 26, 1874, in Fitchburg.

Thus the Commonwealth owned the Western Hemisphere's longest tunnel (until 1916), and according to the state auditor—its true cost will never be known—paid $14,034,184.68 plus interest, or $17,332,019.57 for it.

A quarter-century of tunneling was complete, with a new rule book for both mining and politicking as a result.

1874: *West Portal facade built, track laid.* **1875:** *February 9, first train; April 5, first freight; October 13, first passenger train.* **1876:** *July 1, Massachusetts officially accepts Hoosac Tunnel as completed and opens it for traffic.* **1877:** *East Portal facade built.* **1879:** *Boston, Hoosac Tunnel & Western RR opens, North Adams to Rotterdam, N.Y.* **1887:** *Fitchburg RR buys tunnel and T&G, absorbs VS, T&B, V&M, and controls B, HT&W.* **1899:** *Steam-powered fan installed at Central Shaft.* **1900:** *Boston & Maine RR absorbs Fitchburg RR.* **1901:** *B&M tries oil-burners at Hoosac.* **1910:** *Electrification started; B&M controlled by New Haven RR.* **1911:** *May 18, first electric operation; Central Shaft fan electrified; block signals installed throughout bore.* **1926:** *Tunnel resignaled and deepened at West End.* **1942:** *Two New Haven RR electrics bought for World War II traffic.* **1946:** *August 23, diesels kill electrified zone; CTC signal system from Williamstown to Soapstone opened.* **1957:** *Hoosac single-tracked for piggyback flat car clearance.* **1958:** *November 30, last passenger train through tunnel.* **1973:** *Welded rail installed; track centered for tri-level autorack clearance.*

CHAPTER  A CENTURY OF OPERATIONS

The Shanlys had left Massachusetts with a tunnel, period. 1874 was spent with N. C. Munson's crews grading and laying track while contractor C. McClallan & Son both quarried and erected a gneissic facade on West Portal. The stone came from the J. L. Bassett farm in Northfield, Mass., and the facade added 50 feet, bringing Hoosac's total length up to 25,081 feet, or 4.7501 miles.

At last, on February 9, 1875, the first train gingerly poked its nose into the "Great Bore" at 3:05 P.M. and emerged 34

First locomotive through "the Great Bore," February 9, 1875.
Author's collection

minutes later. The westbound "N.C. Munson"—named for its owner—hauled 125 people aboard three flatcars and a boxcar with George Cheney at the throttle. The cab was packed with Cheney's two sons, Consulting Engineer Thomas Doane (back from the Burlington to supervise the final engineering work), his son, Treasurer Auston Bond, Chief Engineer C.P. Granger, N.C. Munson, B.N. Farren, and one other soul. To commemorate the historic occasion, the locomotive was promptly renamed the "Columbus."

On April 5th, the engine "Deerfield" hauled the first freight through Hoosac with two carloads of local freight and twenty carloads of grain from O. Boutwell & Son, Troy—destined for J. Cushing & Co., Fitchburg. Appropriately, the first westbound freight carried Burleigh Rock Drills.

Governor Gadston and his officials had to view the state-owned project, so they arrived October 13th on the first passenger train. The little 4-4-0 steamed into North Adams decorated with bunting, temporarily renamed "Governor Gadston," and even sported a picture of "His Excellency" under the headlight.

B. N. Farren did the final work on Hoosac when he began in February, 1875 to widen its tight spots, install brick arch over any weak points, and lay a flagstone drainage system.

All trains through Hoosac were his or Munson's responsibility until Massachusetts accepted Hoosac as completed on July 1, 1876. Only their locos were used and only after a toll was received!

From 1876 to 1885 Hoosac Tunnel and T&G were operated by a state manager who levied a user toll on the bore and up to three miles of T&G trackage for access. Tolls were received, but maintenance costs were higher. He supervised the building of a 39½ feet high, 47 feet wide $12,000 East Portal facade between August and November, 1877, and double-tracking in 1881.

Since Hoosac was a "public tunnel," a railroad was proposed from Boston to North Adams via Northampton. At North Adams, the line would branch off to Buffalo or Oswego, New York. The Boston-Northampton and North Adams-Rotterdam, New York links were built, but the master plan died with its promoter. The North Adams-Rotterdam link, the Boston, Hoosac Tunnel & Western R.R., was a godsend to the "Hoosac Tunnel Route." This road connected with the New York, West Shore & Buffalo, also built on the deceased promoter's survey, and enabled the line to bypass the Vanderbilt-controlled New York Central & Hudson River R.R. Without this, for example, no Pullman Sleeping Cars could have been operated into Boston, for the Vanderbilt Lines would handle only their Wagner Palace Cars.

During the early '80s, Massachusetts settled its bills with the Shanlys, and Hermann Haupt! With Haupt they had no choice, since until Massachusetts bought the 12,372 shares of "worthless" T&G stock H. Haupt & Co. owned, the T&G and Hoosac could not be sold to the Fitchburg RR. Haupt finally settled for $150,608 after a legislative fight lasting a decade and led by foe turned close friend—Frank Bird! The final twist to Haupt's $100,000 loss at Hoosac. The Shanlys received about $500,000 for the useless central shaft pumps and the repairs made to Farren's completed brickwork after the flood. Neither contractor returned to the Bay State again.

The Fitchburg R.R. finally gained control of the T&B, VS, and V&M in 1885; absorbing them completely two years later. The Commonwealth was also quite willing to part with

Hoosac and T&G, so on February 11, 1887, it exchanged the "blasted bore" for $5,000,000 in 3% Fitchburg R.R. bonds and 50,000 shares of its common stock. The railroad was overjoyed with the deal, for its common stock was worth only $20 a share, so it had only paid $6,000,000 for the package. It also acquired stock control of the B HT & W with total absorption coming in 1892.

The Fitchburg R.R. decided that electric lights would make Hoosac safer and more pleasing. They installed 1300 lamps, one on each side, and 38½ feet apart. The continual water seepage often shorted out the wiring, and they were removed by 1889.

Hoosac was never clear of the smoke and gas from the 4-4-0's and 2-6-0's. To remove this pollution a steam-driven 16-foot fan was mounted at Central Shaft, and the bottom of the shaft was enlarged. Brick arching the diameter of the tunnel was installed with duct openings at track level, and a small room for the trackwalker was hollowed out: the "Hoosac Hotel." It was hoped that this design would clear the air at the tunnel floor, rather than at the ceiling, but still Hoosac was choked with fumes.

Ventilation became so poor that the only way a track-walker could tell a train was approaching was by his lamp's flare away from the approaching engine. Crews often had to lie down on the cab floor to find air pure enough to breathe, as the firebox inferno died to a dull red glow. The only way a crew could tell if they were moving was to stick a broom handle against the rock wall.

Rear-end collisions were the result of pitch-blackness, brain-numbing coal gas, and 85 to 90 daily trains. One in 1892 left four men dead and another the following year killed one. Tunnel workmen were killed or maimed by trains they couldn't see.

On July 1, 1900, the Boston & Maine R.R. absorbed the Fitchburg and by December, 1901, had New England's first oilburners at Hoosac. Ten-wheelers, #1067-75, and 4-8-0 #1083 had been rebuilt by B&M's Keene, N.H., shops with 750-gallon tenders and got poor mileage: 10 gallons per mile! The oil, pumped into the firebox under pressure, was like

Contemporary artwork took considerable liberties. East Portal is highly stylized, and "Central Shaft Station"—which was to have served the population of a village on the surface, 1,000 feet up—never materialized. *Win Nowell collection*

Below: With whistle screaming, an eastbound passenger job emerges into daylight at East Portal. Circa 1900. *Walker Transportation collection; Beverly (Massachusetts) Historical Society*

Fitchburg Railroad---Hoosac Tunnel Route.

CONDENSED LOCAL TIME TABLE.

6-5 '99

Albany and Troy to North Adams and Boston.

STATIONS.	*No. 6	No. 14	No. 4	No. 8	*No. 2	No. 20	No. 28	No. 10	No. 26	No. 30	No. 34	*No. 32	No. 12	§No150	§No158
Albany (Maiden Lane Station). Lv	......	......	......	10A30	......	1P10	......	......	......	......	......	......	......	......	......
Troy "	From Rotterdam.	......	7A45	11 00	......	1 35	12P55	2P35	4P15	5P00	6P15	*11P30	†11P35	......	§9A00
Lansingburgh "		......	7 54	......	......	......	......	2 44	......	5 09	6 24	......	......	......	9 09
Melrose "		......	8 06	11 16	......	......	......	2 59	......	5 21	6 35	......	d11 55	......	9 20
E. Schaghticoke "		......	8 14	......	......	......	......	3 06	......	5 29	6 45	......	d12A06	......	9 28
Valley Falls "	......	......	8 18	11 25	......	......	......	3 10	......	5 34	6 50	......	d12 07	......	9 32
Johnsonville "	*4A10	......	8 28	11 35	*12P15	2 02	2 23	3 17	4 42	5 41	6P57	......	12 16	......	9 38
✠Buskirk's "	......	......	8 39	......	......	......	......	3 27	......	5 50	......	......	d12 26	......	9 49
Eagle Bridge "	......	......	8 44	11 48	12 27	......	2 35	3 31	4 54	5 55	......	......	d12 31	......	9 54
Hoosick Junc. "	......	......	c	......	......	......	2P43	......	C	6 08	......	*12A19	......	......	C
Hoosick Falls "	G4 32	......	8 57	11 56	12 36	2 22	......	3 40	5 06	6 14	......		12 43	......	10 06
Hoosick "	......	......	9 04	......	......	......	......	3 47	5 12	6 20	......	......	d12 50	......	10 12
Petersb'h Junc. "	......	......	9 09	......	......	......	......	3 51	5 16	6 25	......	......	d12 55	......	10 16
North Pownal "	......	......	9 18	......	......	......	......	3 59	......	6 33	......	......	d 1 03	......	10 25
Pownal "	......	......	9 24	......	......	......	......	4 04	......	6 38	......	......	d1 09	Lv.	10 31
Williamstown "	5 07	7A10	9 39	12P26	1 05	2 50	......	4 19	5 38	6 49	......	......	¶1 26	§1P48	10 43
Blackinton "	......	7 14	9 44	......	......	......	......	4 22	5 42	6 52	......	......	......	1 51	10 47
Greylock "	......	7 17	‡9 47	......	......	......	......	4 25	‡5 44	6 55	......	......	......	1 54	10 50
North Adams Ar	5 18	7 23	9 53	12 39	1 15	3 00	......	4 31	5P50	7 00	......	......	¶1 37	2 00	§10A55
Shelburne Falls "	6 04	8 18	10 37	1 15	1 58	3 39	......	5 18	......	7 47	......	......	¶2 25	2 53	......
Greenfield "	‖6 30	8 44	‖11 02	‖1 34	‖2 22	4 02	......	‖5 45	......	8P15	......	......	¶2 51	3 23	......
Worcester "	......	......	1P57	5 00	......	6 55	......	......	......	......	......	......	†6 35	......	......
Fitchburg "	8 50	11 02	1 30	3 32	4 33	5 45	......	8 10	......	......	......	......	¶5 10	5 25	......
Ayer "	9 19	11 32	1 58	3 58	4 57	6 08	......	8 43	......	......	......	......	¶5 48	6 01	......
Boston Ar	*10A20	12P40	3P00	4P50	*5P50	7P05	......	10P00	......	......	......	......	¶7A00	§7P20	......

Boston and North Adams to Troy and Albany.

STATIONS.	*No. 31	No. 11	No. 27	No. 25	No. 9	No. 29	*No. 1	No. 35	No. 7	*No. 3	No. 5	§No157	§No153		
Boston Lv	......	†11P20	......	......	6A45	9A30	*11A00	......	11A35	*3P00	†8P00	§9A00	......	......	......
Ayer "	From Montreal via B. & R. R'y.	¶12A37	......	......	8 08	10 25	11 55	......	‖12P58	3 59	9 01	10 34	......	......	......
Fitchburg "		1 15	......	6A45	8 44	10 50	12P20	......	1 31	4 27	9 30	11 07	......	......	......
Worcester "		......	......	......	8 10	......	......	......	11A45	8 45	8 45	......	......	......	......
Greenfield "		8 22	6A30	8 41	10 55	‖12P49	‖2 02	......	3P40	‖6 47	11 31	‖12P23	......	......	......
Shelburne Falls "		‡3 56	6 56	9 06	11 22	1 12	2 25	......	4 09	7 17	12 00	1 51	......	......	......
North Adams "		5 50	7 45	9 51	‖12P20	1 53	3 04	......	5 00	8 04	12A53	2 39	§7P40	......	......
Greylock "		5 54	E	‡9 56	12 24	......	......	......	5 05	......	......	2 43	7 44	......	......
Blackinton "		5 56	7 51	‡9 58	12 26	......	......	......	5 08	......	......	2 45	7 46	......	......
Williamstown "		6 01	7 55	10 04	12 33	2 03	3 14	......	5 16	8 16	1 08	2 53	7 51	......	......
Pownal "		6 09	8 03	10 10	12 42	......	......	......	5 24	......	d1 16	3 01	8 00	......	......
North Pownal "		6 14	8 08	10 15	12 48	......	......	......	5 29	......	d1 20	3 06	8 06	......	......
Petersburgh Junc. "		6 23	8 14	10 23	12 57	......	......	......	5 35	......	......	3 14	8 15	......	......
Hoosick "		6 27	8 17	10 27	1 01	......	......	......	5 39	......	......	3 18	8 19	......	......
Hoosick Falls "	......	6 34	8 24	10 34	1 07	2 25	3 36	......	5 45	8 44	1A37	3 25	8 26	......	......
Hoosick Junction "	J	6 38	8 32	10 38	1 13	2 29	......	3P56	5 54	......	......	3 29	8 30	......	......
Eagle Bridge "	......	6 44	8 38	10 44	1 18	......	......	......	5 59	B	......	3 35	8 36	......	......
✠East Buskirk's "	......	6 49	E	10 48	‡1 21	......	......	......	6 05	......	......	3 40	8 40	......	......
Johnsonville "	2A18	7 00	8 51	10 57	1 30	2 46	3 55	4 13	6 19	9 05	¶1 57	3 55	§8P50	......	......
Valley Falls "	......	7 06	8 56	11 02	1 35	......	......	......	6 24	‡9 10	......	4 00	Connects at Johnsonville with No. 2.	......	......
East Schaghticoke "	......	7 10	E	11 06	1 38	......	......	......	6 28	......	......	4 04		......	......
Melrose "	......	7 18	9 05	11 16	1 50	......	......	......	6 35	‡9 20	......	4 12		......	......
Lansingburgh "	......	7 29	E	11 27	d2 03	......	......	......	6 45	d9 29	......	4 22		......	......
Troy Ar	*2A45	¶7 38	9A20	11A35	2P10	3 15	*4 45	4P45	6P55	*9P37	......	§4P30		......	......
Albany (Maiden Lane Station). Ar	......	¶8A05	......	......	......	3P40	......	......	......	......	......	......	......	......	......

Time from noon till midnight is shown in heavy-faced figures. All trains run daily except Sunday unless otherwise designated.
* Runs Daily. ‡ Stops to discharge or on signal to take passengers. § Runs Sundays only. † Daily except Sunday.
¶ Runs Daily except Mondays. ‖ Stops for lunch or meals.
A On Saratoga Branch runs only July 11th to Sept. 4th inclusive.
B Stops on signal to receive passengers for points west of Albany or Rotterdam Junction.
C Will stop at Hoosick Junction cross-over to connect with Bennington branch train.
D Stops only to discharge passengers.
E Stops only on Mondays.
F On Saratoga Branch runs only July 10th to Sept. 2d inclusive.
G Regular stop on Mondays. On other days stops to discharge passengers from west of Rotterdam Junction.
J Train from Montreal. Does not stop at Hoosick Junction.
For time of trains at local stations east of North Adams, see regular local folder time table.

✠ E. Buskirks and Buskirks are stations in the same town, about one-half mile apart; west-bound passengers will take trains at the former, and east-bound passengers at the latter.

FOR PARLOR AND SLEEPING CAR SERVICE SEE INSIDE PAGE.

From "Hoosac Tunnel Route" timetable (Fitchburg RR) dated June 5, 1899.
Ruth B. Browne collection, North Adams Public Library

thin lubricating oil. They gave all coalburners a boost so neither locomotive would have to work hard. Their blue oil smoke was worse than coal smoke, so crews still carried wet towels to try to breathe through inside Hoosac.

The soot and cinder accumulation was so great that an annual cleaning was necessary. During the winter, about twenty men would stand on staging in hopper cars and use coarse brooms to sweep off nearly 40 carloads of soot.

To handle the still-increasing traffic, Alco built four 2-6-6-2 compound mallets, #1291-4, and delivered them in early 1910. "New England's Largest" oilburners allowed the 2-8-0's to be retired and reconverted to coal. The mallets were tremendous haulers, but added the problem of superheated steam to the oil smoke, turning Hoosac into a hellhole.

At this time, the Charles P. Mellen management of the New Haven R.R. gained control of the B&M. They had had excellent success with their electrification of the New Haven main line between Woodlawn, N.Y., and Stamford, Conn., so they quickly ordered Hoosac's electrification. It would eliminate the smoke problem and speed the 48,000 cars monthly through Hoosac.

Westinghouse Electric supplied the electrical components while Baldwin Locomotive Works constructed the carbodies for five engines for the electric zone. They were copied from New Haven engine #071's plans and arrived with two geared for passenger service and three for freight. The construction of catenary, substations, and transmission lines went to the Fred T. Ley Construction Co., Springfield, Mass., and started November 1, 1910. The New Haven's Engineering and Electrical Engineering Departments supervised all construction details and developed operating policy. The electric engine crews were taken at company expense to Stamford, Conn., for three weeks training on New Haven engines.

On May 11, 1911, all construction was completed and the catenary energized for the first time. On May 18th, a light engine ran from North Adams to Hoosac Tunnel Station, returned at 1:30 a.m., and took the first electrically-hauled passenger train through Hoosac three hours later. Operations

Catenary crew takes a photo break in the bowels of Hoosac, sometime during the winter of 1910-1911. *Peter Trabold photo; Randy Trabold*

Opposite page: **The interior of the bore at East Portal, lit by the reflection of sunlight on snow. The distant pinprick of light is a signal.** *Photographed by Peter Trabold in 1930; courtesy of Randy Trabold, North Adams Transcript*

officially began on May 27, 1911, with the Governor's signature on a bill allowing the Berkshire Street Railway—also owned by the New Haven—to sell power to the B&M for their operation.

With the #5000s hauling all steam locomotives with banked fires, soon it was not unusual to see from Central Shaft out either portal: 2.37 miles. It was proposed to extend the zone from East Deerfield yard to Williamstown, but the B&M's independence and the realization that the existing motors weren't powerful enough for the job killed the dream.

The electrics' enginemen soon suggested improvements for their locos. The headlights went under the engineer's window, the steel tube pilot was replaced by two running boards, and a catwalk was built so traveling between the rocking units would be safer. However, management refused to replace the 110-volt headlight and kerosene marker lights with a system from the 32-volt tap on the main transformer. One day the headlight failed and a brakeman stuck his head out the cab door to check if the pantographs were unlocked and up. With a "whoosh," another train rushed by in the

bore, taking the man's head with it. Quietly, a complete electric lighting system was quickly installed in all five units.

Even with electric operation, smoke occasionally filled Hoosac. On February 21, 1912, eastbound freight Extra behind Pacific #3633 was waiting to move into the East End yard when Passenger Train #2 behind Electric #5004 ran into its buggy about 1000 feet inside East Portal. Between the 11,000-volt catenary and the overturned caboose potbelly stove, 17 freight cars and the electric were soon blazing. Four men lay dead: Engineer Archie Simonds, Assistant Engineer Henry Greig, Apprentice Engineer Luther Davis, and the freight's flagman, Ruben Kent. Kent had been sent back to warn the oncoming passenger train and had been picked up by it, for his few remains were found in the engine wreckage with those of the other men. Archie Simonds had been ill, and his assistant had been running the motor. They probably saw nothing in the blackness as the caboose crashed over the engine frame, crushing the men between the front wall of the unit and the transformer-room partition, killing them instantly.

The steamer's engineer went forward to help the men, but heard nothing, so he tried to haul the electric away from the burning caboose, but the brakes were frozen. All he could do was to back his train to North Adams. The fire was fanned into blast-furnace intensity by the constant draft through Hoosac. It took four days to quench the fire and remove the bodies and wreckage.

A faulty signal was blamed for the accident, but Signal Maintainer William Lasher testified that his signals worked properly. Finally it was surmised that the train's "red board" was hidden by the dense smoke, and the crew was still looking for it when the train crashed.

By 1913, traffic was so heavy—70,000 cars monthly—that the Berkshire Street Railway's Zylonite powerhouse couldn't meet the demand with its 6000KW generator. Power was then taken from the New England Power System's plant in Vermont, and transmitted to the uncompleted "#5" station three miles above East Portal. There motor-generator-type frequency changers converted the 60-cycle AC to 25-cycle and then sent it down to the East End switching station or over to Zylonite. By 1915, all power for the zone was coming from three railroad-owned generators at "#5" totaling 8, 600KVA with Zylonite for switching and stand-by service only.

Another rear-end collision in Hoosac on June 9, 1915 sent 29 carloads of perishables and livestock up in smoke, but no lives were lost. A month later that troublesome brook over West End flooded, and operations were suspended while men worked like ants digging away the dirt that filled Hoosac nearly to its roof.

Two years later, B&M was back at Baldwin-Westinghouse for two more engines to help handle the wartime traffic. Because of increasing train speeds, these engines came with passenger-speed gearing and the three older freight engines were regeared, giving all seven locomotives the same power specifications. The railroad west of Greenfield became the Berkshire Division on May 28, 1917.

By 1920, the big #3000 class 2-10-2's were handling most of the Berkshire Division traffic, and two electrics would be on the long trains through Hoosac. With these locos, clearance was so critical in Hoosac that the side windbreakers had to be closed before entering, or they would be swept away when another train was met inside. 3000 feet of the West End was deepened 18 inches during the last half of 1926, but the side clearance was never remedied.

By the late 20s, the #5000's were being overhauled and modified to handle the coming Lima Superpower 2-8-4's, 4-6-2's, and the Baldwin 4-8-2's. New pantographs with two pressure points against the wire were installed, and the air reservoirs were grouped between the pantographs rather than beside them. 11,000-volt to 368-volt transformers of 1350KVA rating were installed as well as Sprague multiple-unit controls that allowed one crew to operate two, three, or four units together. The first of the famous square B&M heralds went on the electric's dark green sides. And the short-lived Berkshire Division faded back into the Fitchburg Division.

That shiny copper catenary wire was lethal to anyone or thing that came too near it. It was an interesting pastime during slack moments in North Adams yard to watch pigeons self-destruct as one stood on the wire and tried to kiss a mate on the other side of an insulator. A sudden "puff" and a few feathers would float gently down to the ground! Less humorous would be the fate of the hobo on a boxcar top who got less than a foot from the wire or the careless fireman swinging a waterplug to the steamer's tender.

Hoosac was always plagued with an inadequate signal system. Originally there was a tower at each portal and a telegraph connection between the two: the Lock and Block System. No train could enter Hoosac until the previous one had exited. With double-track, two trains could be in the bore together, but only on different tracks. Electrified operation brought the first semaphore system with track blocks through the tunnel. All previous systems had just an annunciating circuit at each portal. If single-track operation was necessary, the "Staff System" was used. Both towers were connected to identical machines that held a slotted staff. By an electrical interlock, only one staff could be missing from the two machines at any given time. This staff, in a rubber pouch, was given to the engineer on the only train passing through the tunnel. It would be placed in the machine at the opposite tower and another staff removed for an opposing move. If the single-track operation was for several days, a signalman would have to open up the East Portal machine, remove some of its staffs, and take them over the mountain, for more traffic moved eastward than westward. Otherwise the West Portal machine would be out of staffs.

In 1926, the Union Switch & Signal Co. completely resignaled from Williamstown to the end of Soapstone Siding—three miles from East Portal. Three-light interlocking signals were connected to the Armstrong interlocking towers at both portals and B&M's first Centralized Traffic Control system was installed at East Portal Tower to handle the switches and approach signals at Soapstone Switch.

Application of Automatic Train Stop to the electrics came at this time. This system energized a coil mounted on the track that induced a current in a pickup device mounted on the side-frame of the lead axle of the electric and immediately applied the airbrakes when the locomotive passed a red signal.

A worn-out main frame caused motor #5006 to be scrapped in February, 1942, and with this power loss and tonnage soaring, B&M quickly looked around for more "juice jacks." They found two New Haven electrics of similar vintage and had #071 and #072 in North Adams for ICC inspection by October. These motors were geared much higher than their B&M counterparts, so #5007 and #5008, as they were renumbered, could really roll the "Varnish." Trips with the "Minute Man" were often made in six minutes instead of the allotted eight, and Signalman Bill Lasher still recalls one scorching trip behind #5007 that took just five minutes, portal-to-portal. Towermen wrote off the extra minutes.

By late 1943, the first of twenty-four two-unit sets of 2700 horsepower EMD FT freight diesels arrived on the B&M. Since they were cleaner and more powerful than even the #4100s, more and more freight went behind them, and they didn't need electrics. On March 17, 1945, motor #5005 crashed into diesel #4212 A+B, which was stopped about

Electric No. 5005 and sister coasting out of West Portal with a 4100 Class 4-8-2 and train in tow. Circa 1940, photographer unknown. *Ruth B. Browne collection, North Adams Public Lib.*

B&M No. 1293, one of the oil-fired 2-6-6-2 compound mallets used in Hoosac during the year prior to electrification. *Walker collection, Beverly Historical Society*

Hoosac Tunnel has tremendous allure for railfans. Here's a well-patronized excursion train of The Railroad Enthusiasts halted at East Portal to allow its riders to trackwalk into Hoosac's inky interior. The date is May 22, 1938. *Photo by Stanley Y. Whitney; author's collection*

Below: **The 1978 image! A new 3000 hp Boston & Maine GP 40-2 on eastbound NE-2 emerges from East Portal into a snow-clad January landscape.** *Carl R. Byron photo*

500' inside East Portal. The 4212's sanders had malfunctioned and the units were sitting on a solid layer of sand, insulating them from the block signal track circuits. A false "high green" resulted and in the smoky darkness #5005 nosed into the diesel at 10 m.p.h. The electric's cast steel truck frames shattered, dooming it to the scrap line for parts, but the diesel was only dented.

Saturday, August 23, 1946, #5004 took Midnight Train #62 and its steamer east to Hoosac Tunnel Station and dead—headed back to North Adams for the last time. The next Friday, the four remaining electrics were coupled together and hauled—by a steam locomotive—to Boston for scrapping the following February. By the middle of September, the wires and lineside equipment was also scrapped. Hoosac added one more "first" to its list: the first mainline tunnel electrification to succumb to the diesel.

With the electric zone abandoned, the signal system was again revamped. General Railway Signal put the trackage from Soapstone Switch to Williamstown under CTC control from a new North Adams tower, about a mile from West Portal.

With diesels and CTC, Hoosac Tunnel was just another five-mile section of the Fitchburg Division. One problem the diesels did present: their fumes clouded the yellow lenses on the tunnel signals, turning them red. Special lenses solved the problem. Also a twin-motored fan was installed at Central Shaft to expel the fumes.

Still, the "Great Bore" was a very uncomfortable place to be alone, especially if you weren't planning on it. Signal Maintainer Bill Lasher often drove his "high-rail" car through Hoosac after a train had just passed. One afternoon as he headed for North Adams and supper, he saw a red lantern in the darkness ahead. Braking to a stop, there was the rear-end flagman from the freight ahead of him. The engineer had taken off, assuming the flagman was aboard. The thoroughly soaked and frightened brakie was so happy to see anybody that, as Bill recounted later, "If I'd been a girl, he'd have kissed me!"

The 1950s saw little done at Hoosac as B&M started sliding downhill. A steel storm door controlled from the tower replaced the hand-winched wooden ones in the fall of 1954 at West Portal. Three years later, the double-track was reduced to single iron 3-feet north of center to give clearance for piggyback flatcars. On November 30, 1958, the last passenger train, Buddliner #6155, left Williamstown for Boston, ending 83 years of through passenger service.

Hoosac was only in the news once during the "60's" when on February 23, 1967, an eastbound freight swung from East Portal across the grade crossing and crashed to a halt with 67 cars derailed around the curve and into Hoosac Mountain. Hoosac had become the "boxcar mine" long before that "service interruption" was over after working 'round the clock for eight days.

With passenger service history, let us board a westbound freight extra at East Deerfield yards, courtesy of the B&M operating department and check out what remains of the T&G and its tunnel.

Swinging aboard the locomotive, we realize the entire train consists of empty coal cars headed to West Virginia after leaving a load of 10,000 tons of coal at a massive power plant in Bow, N.H. Loaded and ready to move, this train grossed 13,000 tons and was hauled by four GP-40s and an F-7 booster totalling 13,500 horsepower. These five engines have no trouble easing our 3000 tons into motion as the signal bridge clears. Out of East Deerfield and over the bridge at 15 m.p.h., there are only a couple of hundred amps showing on the ammeter. Crossing under the Connecticut River main line—Springfield, Mass., to White River Jct., Vt.—the throttle inches out another notch or two, and the five V-16s lift our speed to 30 m.p.h. Grass and cracked concrete sweep by, marking the site of the demolished Greenfield Depot. Over the Green River we roar; what would Hermann Haupt think of this bridge and train?

"High Green" at West Deerfield interlocking, and those 20 traction motors propel us effortlessly up the single track fastened precariously between the mountainside and the Deerfield River. Soon we pass South Bridge Station where the trolley line connected the B&M to Conway from 1890 to 1920. "Wah—wah—wa—wa," across Bardwell's Ferry Road onto another bridge, and we're snaking up the opposite side of the Deerfield.

Here the New Haven R.R. once connected with the B&M via the New Haven & Northampton R.R., which completed trackage from Northampton to Shelburne Jct., so called, in 1881, and then had trackage rights into Shelburne Falls. At its peak of operation, the junction boasted an Armstrong Interlocking tower, complete with pipe-thrown switches, and a passenger station. Today only a tree-covered right-of-way and some scattered granite blocks remain to mark the spot.

The Shelburne Falls Depot stands in broken-windowed ruin as we rattle past. No more will passengers purchase tickets for Boston, New York, or Chicago from its fancy ticket booth or relax on the long ornate benches in its matchboard-paneled waiting room for their Pullman berth behind a 3700-class Pacific.

With a steady ½% climb, the diesels easily accelerate to 45 m.p.h., and the rocking hoppers follow like a graceful tail. Soon an eastbound job whips by with the smell of warm brake shoes. Another "Green Board" and we enter single iron and roll through Charlemont. Its station has fared better than Shelburne's, for it is now a boutique and antique shop. The curves grow sharper and the grade slightly steeper.

Zoar Curve, sight of many B&M publicity photos. As the 100 hoppers swing slowly around the giant loop of rail, the sight is impressive as finally the buggy comes into view nearly a mile behind the power.

"Yellow Green" on the Soapstone approach signal. Will there be another freight waiting for us to pass? No, only a ballast train.

The trackside signals are all about 12 years old now, for in 1961 the North Adams Tower was closed and its controls moved first to Greenfield and then—in 1972—to North Billerica, Mass., B&M's main offices. There on a giant CTC board the operator controls every switch and signal from Westminster, Mass., to Johnsonville, New York. It is amazing that this train is showing up as a tiny white light over 100 miles away.

Once there were three tracks at Soapstone, but now an access road has replaced one. The pretty little steeply-roofed Hoosac Tunnel Station is gone, as are the catenary and exchange yard with another of Alvah Crocker's railroads: the Hoosac Tunnel and Wilmington R.R. Originally the "H T &W" reached to Wilmington, Vermont, but was cut back to Readsboro, Vermont, with the completion of Harriman Resevoir in 1923. Now New England Power's Bear Swamp pumped storage project has finished the little road, and nothing remains but a couple of dead-end tracks.

Over the Deerfield by the crumbling remains of Tom Doane's compressor house, past the highway flashers, and into the guts of Hoosac we roll. Windows are slammed shut against the overwhelming roar and the blue fumes visible in the headlight's glare. The hand-drill marks are still prominent in the rock, and overhead the abandoned catenary supports aimlessly hang. Drip—drip—drip—drip—the clear ice-cold water plops on the windshield. After an eternity, three yellow lights come into view—Central Shaft. A quick glimpse of the brick arch and the trackwalker's "Hotel" is all we get. The rock is constantly replaced with brick lining now, and the thought of the manhours those bricks represent staggers the imagination.

The loco's sound and ride changes, for now it's downgrade constantly into Mechanicville, N.Y. A pinprick of light ahead of the headlight bead grows stronger by the minute. Momentarily the brick is replaced by newly installed steel arching that supports the century-old masonry, most of which has stood up remarkably well. With a dazzling flash of light, we burst from West Portal and quietly roll past the abandoned CTC tower and electric engine shop. With brakes squealing, we stop at the North Adams American Legion building—the old depot site.

An hour ago, this train was at East Deerfield yard. It is impossible to guess the number of millions of tons of freight and thousands of passengers a century of operation through Hoosac had seen. High as the price was to the taxpayer, the "Great Bore" has proven itself one of Massachusetts' best investments.

Just a hairsbreadth short of a century after "Columbus," NKP 2-8-4 No. 759 (Lima, 1944) chuffed out of West Portal, the last steam engine to negotiate Hoosac under her own power, en route to head an excursion train out of Hoboken, N.J. The day was Friday, July 6, 1973. *Photo by Don Barbeau*

Five miles of welded 112-lb. rail and a new roadbed centered in the bore with deepened clearances for tri-level autoracks has been completed. More concrete and steel liner plate to shore up some of the brick arch—a 12' by 15' section of which collapsed on August 5, 1972, shutting Hoosac down for several days, has also been installed. Perhaps Hoosac has many thousands of trains left to run through it yet, and with no clickity-clacks to disturb the ghosts of its past.

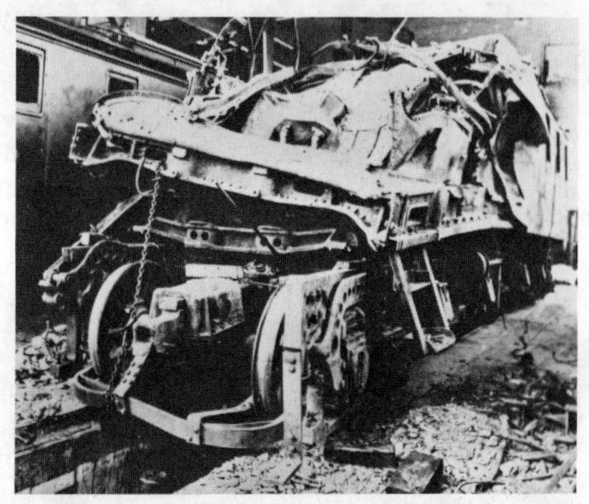

Battered No. 5004 in the electric shop following the 1912 wreck. Amazingly, she was rebuilt and was destined to be the last motor to run in 1946. *William Fletcher collection*

Hoosac's history would not be complete without her spirits. As early as 1865, the workmen were convinced that they were not always alone. On the afternoon of March 20, 1865, a premature blast detonated by Ringo Kelley crushed fellow blasters Ned Brinkman and Billy Nash under tons of rock. Kelley immediately left the locale as fellow workers swore that the spirits of Nash and Brinkman were wandering through the bore waiting to even the score. Naturally, a year later to the day, Kelley's strangled body was found on the exact spot where Nash and Brinkman died. Deputy Sheriff Charles F. Gibson fixed the time of death between midnight and 3 a.m., but no weapon, footprints, or suspects were ever found.

The next brush with the supernatural came after the Central Shaft disaster. All during the following winter, villagers told of vague shapes and muffled cries near the water-filled pit. In the midst of snowstorms or heavy fog, workmen claimed to have seen others carrying shovels and pickaxes appear but for a moment, only to vanish, leaving no footprints in the snow or replies to their calls. These strange visitations on the mountaintop subsided only when all bodies had been removed from the Central Shaft and given a "decent" burial.

Even more reliable than these legends are the recorded accounts of Mechanical Engineer Paul Travers and Dr. Clifford J. Owens—a guest of Drilling Superintendent James R. McKinstrey. Travers penned the following letter on September 8, 1868:

> Last night, Mr. Dunn and I entered the great tunnel at exactly 9 p.m. We traveled about 2 miles. . .(and) in the cold silence we both heard what truly sounded like a man groaning out in pain. . .Yet, when we turned up the

wicks on our lamps, there were no other human beings in the shaft except. . .(us). . .Perhaps Nash or Brinkman —I wonder."

Dr. Owens wrote:

> On the night of June 25, 1872, James McKinstrey and I entered the great excavation at precisely 11:30 p.m. We had traveled about 2 miles into the shaft when we halted to rest. Except for the dim, smoky light cast by our lamps, the place was as cold and dark as a tomb. . . .Suddenly I heard a strange mournful sound. The next thing I saw was a dim light coming from. . .a westerly direction. At first, I believed it was a workman with a lantern. Yet, as the light drew closer, it took on a strange blue color and. . .the form of a human being without a head. The light seemed to be floating along about a foot or two above the tunnel floor. . .The headless form came so close that I could have reached out and touched it, but I was too terrified to move.
> For what seemed like an eternity, McKinstrey and I stood there gaping at the headless thing like two wooden Indians. The blue light remained motionless for a few seconds as if it were actually looking us over, then floated off toward the East End of the shaft and vanished. . .

The blue light form in Hoosac could have been caused by gasses released during blasting and ignited spontaneously. Likewise, the moans and groans may have been the result of unusual acoustical properties in the tunnel. Fog and snow can easily create weird shapes for our imaginations.

Before you decide, visit Hoosac any evening of your choice. When you leave you may not have met the spirits of Nash, Brinkman, Kelley, or some of the other men who died horrible deaths there, but perhaps you will be more inclined to believe that perhaps, even today, their spirits may pass thorugh this history-laden mountain still searching for rest.

Hoosac Tunnel

Commenced: 1851
Length: 25,081'*
East End to Central Shaft: 12,837'
West End to Central Shaft: 12,244'
Depth Central Shaft: 1028'
Size Central Shaft: 15' x 27'
Size West Shaft: 10' x 14'
Height Western Summit: 1718'
Height Eastern Summit: 1429'
Longest Tunnel in North America 1873-1916

Maximum error: 9/16"
Rock excavated: 2,000,000 tons
Height: 20'
Width: 24'
Portals: 766' above sea level
Headings met: November 27, 1873
First train: February 9, 1875
Length brick arch: 7573'
Brick used: 20,000,000+
Cost: $17,332,019.57

Electrification

Contractor: F. T. Ley Construction Co.
Consulting Engineering Firm: L.B. Stilwell Co.

Construction and Maintenance of Zone under New Haven R.R. control (1910-1912)

Electrified Zone

7.92 miles long.
21.31 miles of track total.

Zone sectionalized into 12 parts.

Power: 11,000 volts @ 25 cycles

Construction started Nov. 1, 1910.
First worktrain in Hoosac Nov. 6th.
Two special trains built, complete with oxygen equipment.
Construction completed May 11, 1911.

Zylonite powerhouse coal fired. 6000 KW. 2.42 miles from W. Portal.

#5 Station, New England Power. 3 miles from East End. Waterpowered. 8,600 KVA from 3 generators.

*Or, in the metric future, 7.65 kilometers

Electric No. 2503 (renumbered 5002), built by Baldwin-Westinghouse, 1910. The photo, taken at North Adams, shows the electrics' original appearance. *Walker collection, Beverly Historical Society*

Electric Locomotives

Weight: 262,000 lb
Classification 1-B+B-1
Weight on drivers:
204,000 lb.
Hourly rating: 1352 Hp.
Maximum starting traction:
72,000 lb.
(at 33.4% adhesion factor)
Maximum speed: 37.5 MPH
Hourly rating:
18,480lb traction effort
at 8.3% adhesion
factor at 25 MPH.
Rigid wheelbase: 7 feet
Length between coupler
faces: 48′

Width: 10 feet, 1 inch
Height from rail to locked
down pantograph: 14 feet, ½ in
Driving wheel diameter: 63 in
Idler wheel diameter: 42 in
(on #5008) 36 in
Motor type: Westinghouse 403-A
Drive: geared quill
Gear ratio: 22:19
Delivery date: 4/11
#5000-#5004 Built
10-12/10 arrived as
#2501-#2505 renumbered
immediately
#5005 #5006 12/16
#5007 #5008 (ex-New Haven
#071 #072) 10/42
All scrapped 2/47
(#5006) 2/42

No. 5001 at East Portal, about to walk off with train headed by P-4b Pacific No. 3716, at right, some time in 1938. This is the way the electrics looked after B&M renumbering and modifications recommended by enginemen. *Walker collection, Beverly Historical Society*

Catenary

Outside	Inside
Hung on towers with 150-foot centers.	Hung on bronze hangers drilled and cemented into roof on 100 or 110-foot centers
Contact and Feeder wire 0000 solid copper; hung from 5/8 inch stranded steel cable.	Contact wires two 0000 solid copper hung 5 inches apart. No feeder wire used in tunnel.
Average height above rail: 22 feet.	Average height above rail: 15 feet, 6 inches.

All catenary hung from porcelain insulators tested for 110,000 volts.

ABOUT THE AUTHOR:

"Hoosac Tunnel," writes author Carl R. Byron, "is a well-known engineering feat near my hometown of Colrain, Massachusetts, and many local people have ancestors who worked on the project. I began to research Hoosac out of curiosity and became quite intrigued with the tunnel's history. Gradually, as people requested information on this subject from me, I decided my research was an excellent basis for a book."

After graduation from the regional high school, Mr. Byron attended the Franklin Technical Institute in Boston. He holds a Bachelor of Music Education degree from Lowell State College. He is an avid railroad buff and a member of numerous rail-oriented organizations.